The Gentle KILLERS:

Nuclear Power Stations

The Gentle KILLERS:
Nuclear Power Stations
Ralph Graeub
Translated by Peter Bostock

Abelard-Schuman London

Abelard-Schuman Limited

450 Edgware Road,
London W2 1EG

Kingswood House,
Heath and Reach,
Leighton Buzzard,
Beds.

Printed in Great Britain by
Willmer Brothers Limited, Birkenhead

Contents

Foreword

In the face of the world-wide energy crisis the nuclear power industry will expand enormously as civilization desperately searches for alternatives to the fossil fuels. As oil and coal become scarcer, nuclear power is boosted as the newest and cleanest form of energy available. But is it safe? Its champions insist that it is and, certainly, the atomic industry has a reputation for extreme safety-consciousness. But are the careful safety measures, the controls and the international regulations enough? Ralph Graeub is afraid that they are *not*; that nuclear power stations represent a terrible potential risk in our midst.

The disposal of atomic waste—which cannot be broken down or destroyed organically—presents one problem. A recent report (1973) by the Union of Concerned Scientists in the United States supports Mr Graeub's fears on this score. It says that the techniques for disposal of atomic waste presently being considered by the American Atomic Energy Commission are "dubious in concept, not technically feasible or inadequately supported by technical studies."

The most horrific danger which threatens civilization as a result of the proliferation of nuclear reactors is that of a nuclear accident. *The Gentle Killers* emphasizes that risk is greatest with the American type of water-cooled reactors. The most up-to-date research by prominent American scientists bears this out only too well:

> . . . it is a reasonable conclusion that within ten years or so there may be a catastrophic release of radio-activity from an operating nuclear power reactor. In our opinion, the links in the chain of assurances of reactor safety are substantially defective, and there is presently no action being taken to diminish the risk . . .
>
> Report of the Union of Concerned Scientists.

The American light-water cooled reactors are admitted to be less safe than the British gas-cooled type but Britain is currently considering buying American reactors—for reasons of economy. In the controversy over the wisdom of this policy a consultant physicist has described what could occur if the emergency core-cooling system of one of these reactors were to fail:

> A hundred-odd tons of uranium, intensely radioactive fission products and assorted hardware will slump into a huge boiling puddle that will melt its way down through the steel pressure vessel, the concrete containment and a considerable depth of rock. In the process, the puddle will release to the air an amount of fission products . . . comparable to the fall out from hundreds of Hiroshima bombs.
>
> Amory B. Lovins, *Sunday Times*,
> 25 November 1973.

This sort of total catastrophe is *possible*; smaller, 'less serious' accidents are *likely*. In the face of this kind of danger it is vital that the facts about nuclear power stations should be at the disposal of as many people as possible so that everyone can make up his own mind about them. That is what Ralph Graeub has tried to do in this book.

EDITOR.

Preface

Atomic power stations: yes or no? Everyone has to confront this question sooner or later, and many will feel themselves ill-equipped to answer it. Hardly any other field of technology can point to so many prominent scientists ranged on either side of the argument. Whom should we actually believe?

This book tries to attack the problem from a new angle, by critically examining our present knowledge of the biological effects of radioactivity in relation to nuclear power stations: starting from ecological precepts and relying principally on the publications of the United Nations Scientific Commission on the Effects of Atomic Radiation, UNSCEAR, and of the International Commission for Radiological Protection, ICRP (on whose recommendations radiological protection laws are based).

However, it is not a book only for the specialist. Rather it aims to give the reader of average education a straightforward opportunity to judge for himself the dilemma of the peaceful application of nuclear energy within the framework of other disorders of civilization and the laws of ecology. In trying to achieve this, a certain compromise between a scientifically precise and a generally intelligible presentation cannot always be avoided; but this should have no influence on the formation of an objective, well-founded opinion. The opportunity of forming such

A*

an opinion should be available now to every individual, be he politician, journalist or ordinary citizen.

The ever-increasing manipulation of the radiological factors governing our environment has an immediate effect on each one of us. It must not be that, for lack of adequate information on which to base their judgements, those responsible should have to rely simply on the statements of scientists committed to the progress of technology and the dictates of economics. Such a situation could lead to the wrong decisions being made.

The headlong technological exploitation of inadequate scientific knowledge is the chief cause, not only of the destruction of our environment by pollution of land, water and air, but also of the poisons in our food. New technologies should no longer be "let loose" on society solely on the strength of minority judgements. The estimates of risk and profit usual in many fields of technology can never be applied to the production of energy through nuclear fission, with all its global consequences: because the factors necessary for the calculation of risk lie largely in the dark. The detrimental effects—especially for future generations—cannot possibly be foreseen at the present time.

Neither the scientist nor the expert should therefore make these decisions alone. The "peaceful application of nuclear energy" should also be the concern of politics, morality and conscience. The population at large accordingly has a right to be frankly informed of the dangers and risks it can expect. It is within this context that this book aims to be effective.

Ralph Graeub
DEPARTMENT OF CHEMICAL ENGINEERING
EIDGENÖSSISCHE TECHNISCHE HOCHSCHULE (ZÜRICH).

I · Man ignores
the laws of ecology

1 · INTRODUCTORY

Through technology, science and dynamic industry, one part of mankind has been able to create a unique state of prosperity. The significance and the function of these factors in creating our prosperity cannot be overlooked.

Mankind, however, has been unable to keep up with this rapid development, because legislation, culture and education have lagged far behind. The disadvantages—now impossible to overlook—have only begun to appear in the last three decades. Progress has always been understood as technological advance and the expansion of production, through which the standard of living could be steadily increased. However, all this may be a retrograde development if it proceeds at the cost of our health and the life-support systems of the environment. Mankind has previously paid no attention to this, largely because, as yet, the laws of ecology have been too little developed.

The reality is that mankind is in a position to destroy himself: not only with hydrogen bombs but also quite peacefully with the population explosion, misguided methods of agriculture, the petrol engine, the despoilation of nature, the peaceful use of nuclear fission, et cetera—in short, by the interrelated pollution of land, water and air, through the poisons in our food and through the continuing destruction of our environment and its ecological balances. European Conservation Year in 1970 helped to bring these facts home to a broader cross-section of people and in many political party programmes environmental protection now heads the list.

We must recognize that the dangers of atomic power stations cannot be examined or understood in isolation but only in the context of other disorders inherent in civilization and of the perilous collective situation of present-day mankind. It is a mistake to assume that the risks of atomic power stations can only be judged by specialists and scientists, and that a higher technical education is necessary to do so. Such an opinion is of course gladly offered by interested parties, for it allows a veil to be comfortably drawn over the crucial problems of atomic power stations which no outsider is apparently to be allowed to see.

From time immemorial those who have warned against or opposed contemporary developments have been decried as incompetent laymen, or branded as people who want to stem the tide of human progress. In this way, Nobel Prize winners, doctors, engineers and serious social reformers are all thrown into the same melting pot as fanatics and opportunists. This alone suggests the essential bias of such an argument.

It is very significant that it is often precisely the great scientists who best recognize existing problems. On the other hand, the mediocre talents, technology's hacks, do not usually think very far ahead or reflect whether or not their work is defensible in terms of the general good. They see their responsibility as limited to the job in hand, and would reject as insulting any accusation of immoral practice with respect to the public.

With our God-given reason we can only govern our immediate present with any degree of certainty. Where the future is concerned we generally behave most

2

unreasonably. Not least to blame for this is our archaic, backward-looking, humanistic conception of education, combined with the one-sided analytical thinking of scientists.

In order to understand all these problems, we must first of all make clear to ourselves some of the concepts and correlations of *Ecology*: *the study of the relationships among organisms and between them and their environment.* This is a relatively new science and is not even offered as a school subject in many countries, although an understanding of the present critical state of mankind is hardly possible without it. Present day society cannot therefore be spared the reproach that it is raising its children always to look backwards, instead of preparing them for the present and the future.

According to Professor Dr P. Tschumi (117) it is fundamental to modern ecology to understand that life operates on different *levels of integration*. Five levels can be distinguished:

At the lowest level are the individual *cells*. To this category belong single-celled plants and animals, for example, bacteria.

The next level consists of the multi-celled *organisms*. These are highly organized communities of enormous numbers of different specialized cells. Each single animal, each plant and the individual human being, too, should be included here.

The third level is the *population,* that is, a reproductive society of organisms of the same kind. Because of its heredity it forms a real—if not very noticeable—biological unity. It has its common heredity, its genes. Mankind taken as a whole represents such a population, as do all cats, or all horses; i.e., every species of animals and plants is a population.

But even populations are not autonomous, independent

3

unities. In nature all the plant and animal populations in an area exist in a state of extensive dependency on each other and on their environment, and these living communities form a so-called *ecosystem.* A forest, a lake, every pond, the sea and also every human settlement with its environment are ecosystems.

Lastly, we can discern a fifth and final level of integration of living things, the *biosphere,* the total of all the ecosystems on our planet. This biosphere is also mankind's living-space.

Man must therefore no longer consider himself as isolated from a world which he can alter at will with his science, his technology and his ethic of individualism. Man is much more like plants, animals and micro-organisms, a link in the chain: a population of a living community or of an ecosystem, and therefore subject to the same ecological laws as other living beings—in spite of his supposed freedom.

Every unnatural disturbance of such an ecosystem brings with it weighty consequences, and it is precisely this factor that man has taken far too little into account in his reckless striving after material improvement. It is for this reason that the effects of environmental pollution and the destruction of nature have become so apparent today.

2 · OVERPOPULATION

One of the basic natural laws holds that, in a balanced ecosystem, there are four or more so-called *trophic* levels which have a definite quantitative relationship to one another. They are these:

the *producers,* or green plants;
the *primary consumers,* or plant eaters;
the *secondary consumers,* or meat eaters;

the *decomposers*, for example, bacteria, worms, insect larvae, fungae.

Complex regulatory mechanisms ensure that none of

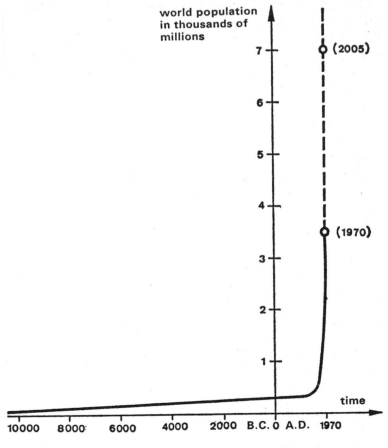

Fig. 1. World population growth over a period of 12,000 years.

these levels grows beyond a certain state. Without this regulation of population density, both the food supply for the population and the decomposers would either be overstrained or destroyed—which would finally lead to a collapse of the population in question, even if it did not bring down the whole ecosystem with it. In the course

5

of their cultural development mankind has gradually eliminated these natural regulatory mechanisms. The progress of medical science has resulted in an excess of births, about which man has concerned himself far too little: so that along with the development of increased life expectancy, the present population explosion came about. The world population now increases at a rate of 190,000 people per day, or 2 people per second (119).

The force of this exponential growth is so explosive that in only 34 years the world population will have doubled. At a similar rate of growth, we would, by the year 3000, have four people per square inch. Obviously this growth must be checked for, in the limited space the world offers, unlimited growth is simply not possible.

Authoritative scientists are of the opinion that there are too many people on the earth even today; and that it is quite inconceivable that underdeveloped nations can ever reach our standard of living. The average American even today consumes ten times more of the non-renewable resources of raw materials than his counterpart in an underdeveloped country. Were the underdeveloped nations to be brought up to the American standard of living, there would be a strain on the environment ten times greater, an unbearable prospect (10). Without concomitant measures for birth control and family planning, the aid programmes in operation today will merely initiate these nations into the worst disorders of civilization.

However, our need for energy is also following a rapid growth rate, and the authorities responsible for the provision of power are zealously calculating the amounts that will be needed in the future. Theoretically, the figures are almost astronomical; nuclear fission is seen as an ideal solution for satisfying these demands. But the adoption of such a solution implies that absolutely no consideration has been given to the inevitable, intolerable disruption of the ecological equilibrium in the biosphere that would result. The exponential growth of population, material standards of living and demands for power will have to be restricted within the next two decades or so if mankind wants to survive. This is unequivocally clear from the results of research by futurologists (E. Basler) (10). According to them, a brake must be put on development within the next 25 years, in order to restore the material balances of the biosphere as an ecosystem.

If mankind had not discovered nuclear fission, other possibilities for energy production would certainly have been developed, perhaps tidal, wind, or solar power stations. There are already concrete proposals for harnessing glacial waters in Greenland. In addition, more work would certainly have been carried out on nuclear fusion, although this, to be sure, does not appear entirely free of problems because of the production of neutrons and tritium that it involves.

Fossil fuels like coal, oil and natural gas will last for the near future at least. The power stations which rely on these are also obviously a blight of civilization, but they are the equivalent of an attack on our health with knives, where nuclear power stations represent an attack with cannons. Thermal power stations could be better designed and constructed without too much trouble. Furthermore the problems of emissions to the air from oil-burning power stations can be controlled in the main (5). However, the site must be chosen carefully, a tall chimney built and either low-sulphur or sulphur-free oil used according to climate. Waste gas filters could also be improved. Obviously such measures will make energy more expensive but this can hardly be avoided if we want clean energy. Economic considerations should no longer take first place. Furthermore, natural gas has just come on the scene, and, although not inexhaustable either, its combustion poses no air purity problems at all.

Without nuclear power stations mankind would certainly not perish. It is entirely wrong to keep quoting Professor G. H. Whipple from Michigan (143), who maintains that nuclear energy represents the cleanest, safest and most sizeable source of energy, and one which has been discovered at exactly the right moment. Pronouncements like this should not be accepted so readily; fission energy does promise mankind further growth at a decisive moment—but against all the laws of ecology. As a result mankind could be prevented from applying in time the necessary braking and regulating measures on material development.

3 · NATURAL SELECTION

If the dangers of nuclear power stations are to be made clear, a further crucial function of ecosystems in nature must be understood. The main aim of these ecosystems is to protect the totality of life. In the course of thousands of years they have established a dynamic state of equilibrium. So, for example, every forest exists in a state of permanent equilibrium even though each tree only lives for a limited period of time. Some encroachment on a particular ecosystem is possible therefore; on a forest, for example, individual trees in it may be felled without destroying the whole forest. In this way the totality of life is preserved. Indeed, it should be clearly understood that the balance of an ecosystem can only be maintained through the intensive weeding out of individual lives. The different levels can only be regulated in this way. Nature is therefore only apparently peaceful. In reality the hard law of "eat and be eaten" reigns supreme, and is the essential prerequisite for a healthy ecosystem and for

8

the maintenance and further development of the species (117).

Through a wise precaution in the natural struggle for existence the weak, the sick and the genetically damaged go under first, and are continually eliminated. So, on the basis of natural selection and under the protection offered by ecosystems, a higher development of the species has become possible: because all that is healthy, strong and well-equipped for life survives and can propagate itself, while all that is weak is eradicated and disappears.

To a large extent mankind has been able to escape from this law of ecology through medical progress and through the ethic of individualism, which emphasizes the well-being, dignity and survival of the individual, without regard for the interests of the population as a whole. (119).

In our present-day society, therefore, everything that is weak or sickly is supported by medical aids, and so can pass on genetic damage. Earlier, all of this was eliminated by genetic disability, in the struggle for existence. Since natural selection no longer operates there is a danger that the human species is becoming weaker in health and degenerating.

It is extraordinarily important that the continual struggle for basic existence that rules in nature is correctly understood. Our entire existence has always been influenced by natural background radioactivity which has been present from time immemorial. However, because of the process of natural selection, it was scarcely possible to transmit the damage caused by such radiation genetically. Today it is acknowledged, even by the ICRP, that the genetically injurious effects of radioactivity begin from a radiation dose of zero (66). In this respect it makes no difference whether it is a question of natural background or artificial radioactivity. It is therefore quite false to maintain that we have always lived unharmed in

9

a sea of radioactivity. This view fails to take account of the original effects of natural selection. Since, however, mankind has sidestepped this process to a great extent with civilization, *even natural radiation is too much*: with the result that any increase in the present level of radiation presents a risk.

The ICRP itself states (58) that for the majority of mankind the risk of natural background radiation is of the sixth order (see page 72), as opposed to a risk of the fifth order for those few population groups living in belts of high natural background radiation: which is accordingly greater! Therefore it is entirely false to continue to maintain that natural background radiation is harmless.

Obviously we do not want to regress to the bygone times of the struggle for a bare existence, and raging epidemics. But we must take all the more care of our hard-won heritage in the knowledge that we have evaded an important ecological law. The release of mutagenic substances into our environment, as has recently been the case from nuclear power stations, must, therefore, be seen as irresponsible. Moreover, as we shall see later, the products of nuclear fission accumulate in the biosphere and in living organisms. With the rapid increase in the number of nuclear power stations the dangers are obviously mounting.

4 · ORGANIC FARMING

Mankind makes further severe and unnatural onslaughts on ecological laws by employing incorrect agricultural methods. In striving to enhance the superficial quality of fruit and agricultural products and to achieve ever higher yields, he pursues chemical pest control and uses massive amounts of artificial fertilizer. In its broadest sense, this includes the use of plant protection agents like insecti-

10

cides (against pests), herbicides (weed killers), fungicides (for fungal or mould diseases) and seed protection agents (against wheat blight, et cetera). These substances can be sprayed, poured, scattered, or vapourized. Although there are precise instructions and regulations for their use, and in spite of official commissions, food laws and inspections, et cetera, governments are not able to prevent our food from being contaminated more and more by the residues of these poisonous products. Apart from that, we are also consuming in increasing amounts other additives, which have been proved to be harmless, such as anti-oxidants, stabilizers, emulsifiers, thickening agents, dyes, flavouring, artificial sweeteners, organic acids, et cetera.

The whole problem of plant-protection agents first came properly under discussion with the appearance in America of *Silent Spring* by the biologist Rachel Carson, who was battling against the systematic slaughter of birds and wild animals. President Kennedy read this stirring book and established an inter-governmental investigation committee which confirmed, in all essential points, the world-wide contamination of biological systems.

The best known instance of this is the global dispersion of DDT, whose good reputation has been almost destroyed. For a long time competent scientists and government experts emphasized that this insecticide was completely harmless for human beings. The warning voices of independent scholars and reformers were either silenced or dismissed as unscientific and untrue.

Too little notice has been taken of the fact that DDT and other chlorinated hydrocarbons are stored in living things and can be concentrated in food chains. Thus, today, almost all marine organisms are extensively contaminated and even Antarctic penguins are showing signs of chlorinated hydrocarbons in increasing amounts. As marine organisms accumulate more DDT the higher they

11

are in the food chain, the marine birds of prey at the end of the chain are most threatened, and it is hard to say whether they are not perhaps already doomed.

The sea as a source of food for mankind is also in peril. Even if every use of DDT were to be forbidden now, we would still have to reckon with its delayed effects on many life-forms in the marine community. Especially problematic is the damage to marine plankton, which must be credited with up to fifty per cent of the oxygen production of the earth.

At the end of every food-chain may appear man, in whose fatty tissues DDT is stored. Thus today every European has an average of 2–3 ppm (parts per million = 1mg/kg body weight) of DDT in his body tissues, every American as much as 12, so that it would no longer be advisable for a cannibal to eat an American! The US Department of Agriculture permits the meat industry only at most 5 ppm. But why 5, and not 0? Because there is hardly any meat in existence that contains absolutely no DDT.

Research in the canton hospital of Aaran (Switzerland) has shown that mothers' milk contained so much DDT that it was no longer fit for consumption. In this field it is still not known exactly what injurious effects DDT and other pesticides can have on human beings. Influences on the nervous system, disturbance of hormonal balance, brain damage, and similar effects are suspected or, in some cases, have been isolated.

In the summer of 1971, the Federal Republic of Germany became the first country to decree that DDT could not be manufactured, imported or exported; both its acquisition and its application—even in the hygiene sector—are forbidden. Even products (meat, ointments, lipsticks and so on) found to contain traces of DDT may no longer be imported.

Although fire and petrol are dangerous even by them-

12

selves, it is only in combination that they cause an explosion. The uncontrollable combined effects of different insecticides and the reaction of other chemicals which contaminate the environment constitute an extremely dangerous, but largely unknown, factor. DDT moreover provides a classic example of the hopelessness of trying to establish tolerance doses for poisons which can accumulate in the body tissues.

All these chemicals are used with official permission even though they are manifestly poisoning us and our environment. It is worth reminding ourselves at this point of the case of lead—added to petrol as an anti-knock agent. Mercury is also worth mentioning. It reaches the rivers and seas in the effluent from cellulose factories, paper mills and plants producing plastics and chlorides. It can even reach the atmosphere through the burning of paper, to which mercury is often applied to prevent fungal growths developing. Sprays containing mercury which are used to kill off injurious fungae are a real source of danger. As is the case with DDT, this metal can only be removed from the body with the greatest difficulty because it accumulates in and then becomes concentrated in organisms. Every ton of mercury that is extracted from a mine remains once and for all in the biosphere.

Newer and newer sources of danger are being discovered all the time. Substances to which far too little attention was paid in the past are accumulating now in the biomass. For example, it was discovered recently that fat-soluble polychlorinated biphenyls (PCBs) have already caused contamination on a world-wide scale. Both human fat and mothers' milk contain even greater amounts of PCBs than of DDT! PCBs are widely used in electro-technology in transformer oil and in insulation materials, and also in the plastics industry, as a softening agent. As PCBs had no connection with pesticides they

13

did not have to undergo testing or sanctioning and passed practically unnoticed—they were certainly not examined for their toxic effects. However, it is now known that they are biologically active, although there is as yet no clear understanding of their toxic properties. They do influence calcium balances to the extent that broken bones heal badly and birds are unable to incubate their eggs properly because the calcium shells are too thin. And no method of biological degradation is known for PCBs. They should therefore be banned unconditionally.

It is probably one of the crassest of errors to allow substances which can build up in the biomass to be freely dispersed in our environment. Official supervision, establishing tolerance levels and passing regulations, just cannot bring them under control, either in the agricultural sphere or when they are discharged as waste in the environment (for example, as radioactive waste from nuclear power stations). In the case of radioactive waste it is naively calculated that the radioactivity will simply be harmlessly absorbed in the natural elements!

Far-sighted scientists condemned the use of poisons in agriculture long ago, and demonstrated that land could be cultivated quite satisfactorily by organic methods without resort to poisons at all. First of all we must decide just what organic farming implies. In order to do this we have to make clear what we mean by the concept of quality in fruit and vegetables. *Quality* implies two main attributes:

Market value: this means fruit and vegetables are judged by external appearance, cosmetically. The consumer normally selects fruit on grounds of appearance, freedom from blemish, and size.

Organic value. People do not ask about this as readily although it actually represents the decisive criterion for judging the quality of produce.

Organic value has four important distinguishing characteristics: *nutritional value; wholesomeness; health value; freedom from toxins.* These four characteristics of produce are all essential to life and health, but the average present-day consumer scarcely considers them. Organic cultivation of land is designed to produce food that has the greatest possible organic value; this implies that considerations of external appearance or market value become correspondingly less important. This kind of cultivation means that it is necessary to activate the land; and this involves, as a priority, not feeding the plants but stimulating the soil. Soil is not the lifeless, inert substance that many people imagine, existing simply to provide plants with mineral salts and give support to their roots. Healthy soil swarms with bacteria, fungae, yeasts, algae, worms, insects and other small organisms. Even the common earthworm is represented by a population of up to 65,000 per hectare (2·5 acres) in healthy soil. Its function is to plough and loosen the soil.

The ecology of healthy soil does not depend only on the presence of certain minerals; it also contains organic substances from decomposed vegetable matter and animal excreta, and countless numbers of organisms all living in the one community. New humus is formed from the organic substances through the action of these soil organisms, and the more humus the soil contains, the healthier it is. Humus takes care of the necessary aeration, increases porousness, stabilizes the soil temperature and provides food for the living creatures in the soil. Soil is only healthy and fruitful if humus, minerals and micro-organisms are all present in an exact balance. Modern over-fertilization with nitrogen-phosphorus potash can destroy the mineral balance in the soil. It is well-known that plants grown in freshly cleared forest land hardly ever suffer from pests in the first years of growth.

Organic farming respects the ecological laws: so no artificial fertilizers or chemical poisons are used, for these would destroy the rich life of the soil. Natural rock powders can be used to compensate for the loss of mineral substances. Mouldy dung or putrid liquids are never used for manure: compost or organic waste products, which have already been through a natural process of decomposition, are substituted. For surface fertilizing even bone and bonemeal may be used. These methods make for a healthy soil.

Plants grown in this kind of soil have extraordinary powers of resistance to pests. The Creator did not put pests and parasites on the earth just to make farming difficult for man. These things should be thought of in the context of natural selection, as inspectors whose job is to weed out the sick and the feeble (117). Today man is breeding "sick" plants in their millions with faulty agricultural methods and vast monocultures which contradict all the ecological laws. The "sickness" may not always be obvious: may even appear insignificant. Because science is unable to demonstrate this mistake with the tools currently at its disposal—and seems totally uninterested in trying to forge new ones—people simply maintain that there is no deterioration taking place in agricultural products. But the increased incidence of diseases caused by pests proves that there soon will be!

Pests find living conditions ideal and multiply in vast numbers. Man retaliates, and combats them with chemical poisons which make the soil and plants even sicker. Because insect generations succeed one another so rapidly, the insects quickly develop resistances to these poisons; and other, much stronger ones, have to be used. The soil and the plants become sicker still. It is a vicious

16

circle from which, on this basis, there seems to be no way out.

Organic farming, on the other hand, does not suffer from pest problems on anything like this scale, Organically-worked land often survives like a protected island in the midst of threatened areas. A classic example of this was demonstrated during the outbreak of foot and mouth disease in Switzerland in 1968, when several hundred organically-worked farms, under the direction of Dr Hans Müller in Grosshöchstetten, remained free of the disease, although they lay right in the middle of the infected area. It is incredible that to this day science has refused to investigate this fact (9). It has long been shown in practice that fodder from organically worked soil makes animals more resistant to epidemic diseases. The same has been observed of human beings whose food is organically grown.

A first-class example of a large operationally-successful organic farm is provided by the Biotta Company, in Tagerwilen (Switzerland), who produce the vegetable and fruit juices of the same name. Thanks to this company's use of a natural process of fermentation the usual addition of preservatives has been avoided.

The total net production of an organically operated farm can never be as great as that of an unhealthy and unstable farm dependent upon the infusion of high-energy phosphate-nitrate manures. But we are not getting our priorities right if we allow a growing and uncontrolled world population to be satisfied with damaging and unnatural farming methods. Instead, population growth must be checked—and as quickly as possible—in the interests of maintaining the health of man and Nature, and to make sure that we have adequate supplies of healthy foodstuffs.

Converting land to organic farming will certainly take several years because it depends on how quickly the ecological balances in the soil can be restored and on whether humus can be re-formed. The soil must also be in good enough condition to be able to break down any remaining poisons.

Restoring a piece of land to full fertility with organic methods is a large and difficult task, for it cannot be assumed automatically that the quick results, possible with chemical fertilizers, will occur. But restoring land does not just mean making one sudden effort: it means building up a healthy soil, a healthy ecosystem. Only those who have really grasped the meaning of the laws of ecology can understand that success depends on cooperation with Nature: and that the opposite will result if man continues to defy natural laws by using poisons. Conversion to organic farming will have to be effected gradually, through an increasing demand on the part of consumers for organically-grown produce. A slight rise in prices must be anticipated; but experience has shown that the enlightened consumer is likely to accept this as reasonable. And many farmers have been surprised at how smooth and successful the conversion to toxin-free cultivation can be, if the right foundations are laid at the beginning.

We must understand the full significance of organic farming before we can fully appreciate the tragedy that threatens us through the discharge of artificial radio-nuclides from nuclear power stations. Up to now it has been possible for organic farming to produce virtually poison-free food. Virtually no impurities are found in organic food, in contrast with produce created with chemical poisons. Toxic traces in organic food are very much smaller than they are in produce that has been sprayed and affected by the already extensive contamination of the biosphere.

The levels of dangerous radioactivity in many foodstuffs were raised everywhere by the fallout from atomic bomb tests. Even if atomic bomb tests were to be stopped now we would still have to reckon on higher than normal concentrations of radioactivity for many years to come. But apart from this there are radionuclides, present in the discharge from modern power reactors, which can become concentrated in plants and animals. Increasing numbers of reactors will certainly cause greater concentrations of radionuclides, particularly in the immediate vicinity of the power stations.

So, although the barrier reef of the atomic bomb tests has barely been navigated, a new source of danger is already looming up, against which organic farming is powerless to defend itself. Dr M. O. Bruker, Medical Superintendent in Lemgo/Lippe, asserts that "the dangers from insecticides are child's play in comparison with those from radioactivity" (20).

The discharge from nuclear power stations is no doubt well supervised; but that supervision is not sufficient. We have enough experience to know that, even with such measures, the gradual contamination of soil, air, water and living creatures cannot be avoided as long as there are substances released into our environment that are difficult to break down organically, or that are completely indestructible (like radioactivity), and can accumulate in food chains. Tolerance doses are absolutely useless against all cumulative poisons.

5 · MEDICINE AND HEALTH

Only those who are not completely dazzled by the wonders of modern medicine can really understand the importance of natural foods for our health. Not that medical advances should be underestimated in the least —we need only remind ourselves of our recognition of the importance of hygiene, the lowering of infant mortality rates, the battle against contagious diseases, the highly refined surgical techniques that now exist, et cetera. We do not want to make the terrible mistake of condemning these positive results simply because they appear to conflict with some of the ecological laws. In the last analysis we are *human* beings, and we do not want to return to a primitive state of ruthless natural selection: but, as Dr G. Schnitzer says (98), we ought to be seeking a compromise between the demands of civilization and health.

Civilization does not automatically imply a degeneration in health standards. But our health *has* deteriorated because we have clung to a number of deeply ingrained habits. We could change this situation if our environment were not subjected to constant poisoning with chemical and radioactive substances.

More and more progressive and far-sighted doctors are coming to realize that medicine's most important task lies, not solely in the treatment of illness, but in the maintenance of health—as advocated by practitioners of natural healing for many years. Health cannot just be taken for granted, it must be cultivated by leading a healthy life, which includes good nutrition. So, contrary to present practice, medical science should primarily seek the fundamental precepts for the maintenance of health, and only when these have been established should it treat illness as such.

Health simply cannot be bought at the chemists. But most people think it can, and this has led to an unprecedented abuse of medicaments. It may sound like a contradiction, but it is true to say that the act of healing, which should benefit man, has almost become a danger to him. Dr David Spain (104), looking through a weekly medical journal, discovered that articles about iatrogenic ills—those caused by doctors—had doubled over a period of ten years. These ills fall into three main categories, concerned with: 1. pernicious side effects of chemical medicines; 2. delayed damage after X-ray treatment, either for diagnosis or for therapy; 3. damage inflicted on patients through taking tissue samples or through injections. Spain also contends that it could be shown at any given time that out of 1,000 patients in a hospital, 50 would be suffering from complications arising from chemotherapeutic measures. But health cannot be bought for money from expensive doctors and at expensive clinics. Nor is it a kind of capital which can be frittered away for a whole lifetime; it demands constant care and attention.

Nevertheless modern man—mainly because of inadequate health education—does not understand that the maintenance or restoration of health is entirely in his own hands and has to be paid for by a number of (apparent) sacrifices. Illness arises most frequently when people do not pay constant attention to the normal functioning of the body.

Unfortunately, professional medicine pays far too little attention to these facts and occupies itself one-sidedly with the diagnosis and treatment of illness. Even so-called preventive medicine, which should actually be concerned with the conditions necessary for healthy functioning, or general well-being, only abandons this obsession with diagnosis and treatment with the greatest reluctance.

Without doubt, therefore, the modern stress on the importance of the *symptoms* of disease takes far too little account of the actual *causes* of illness. These causes originate primarily in erroneous ideas about life and nutrition, from the widespread destruction of the environment and from its gradual contamination. Yet, not only do these causes of illness pass largely unnoticed, but they are actually becoming more deeply entrenched. Without a healthy environment the good health of human beings can only be preserved to a very limited extent—and healthy food cannot be produced at all.

There is therefore no point in everybody rushing off to become fanatical vegetarians and apostles of health cults.

In industrialized countries there is no guarantee at all that the citizens' health will be protected, because the state does not really concern itself with promoting healthy nutrition. On the contrary, official departments, either from ignorance or from lack of autonomy, authorize farming methods, food and eating habits which are injurious to health. Living according to these bad eating habits encourages the food industry to turn out a variety of products which are designed to tempt the customer rather than provide him with good nutritional value. The public is never taught about this state of affairs.

It is therefore crucial for every citizen to try to solve the question of nutrition for himself and his family—and indeed to do it in the face of the customary, traditional, so-called respectable, middle-class eating habits of the present day. A widely publicized account of natural healing would help him.

Naturally not all of this should be accepted uncritically. Many reformers hold quite contradictory views, and all the extremes should be avoided. The best sources of information are medically trained doctors who have proved in the course of their practice that greater and more lasting success can be achieved by combining natural healing methods with their own purely medical knowledge. The result of this combination of professional medical knowledge and natural healing methods would mean the spontaneous growth of proper preventive behaviour in health care. It is particularly disturbing to note, of course, that there are charlatans active even in the fields of natural healing and modern nutrition, who propagate false and often dangerous theories. Critical examination of the evidence therefore cannot be too strongly advocated.

Dr Maximilian Bircher-Benner (1867-1939) must be counted the pioneer of modern health food science, for he was the first to move away from purely mechanical ways of thinking about life and nutrition. For him, natural raw foods rich in vitamins were a priority. Although medicine at that time knew nothing whatsoever about the vitamins to be found in raw foods, Dr Bircher found the correct path instinctively and had quite sensational success as a result. "Birchermuesli," made of fruit,cereals and milk, is well known. His son, Dr Ralph Bircher, now over 70, has continued his father's work in Zürich (at the Bircher-Benner Clinic). He has achieved and is achieving extraordinary feats for modern nutrition. He is kept particularly busy as a publicist and has made it his special concern to select and organize some of the many findings that have emerged from current international research in nutrition. For many years his work has been published in *Wendepunkt* ("Turning Point").

It is staggering to discover how much research, all over the world, confirms the validity of Dr Bircher's modern, natural and essentially lacto-vegetable dietary theory, although professional, academic medicine takes little notice of these findings.

Recent research by Dr Ernst Kofrany (17) at the Max Planck Institute indicates that the currently accepted protein theory, which holds a central position in orthodox dietary science, is incorrect. It implies that the need for protein is not satisfied by the valence[1] of individual

[1] Valence: biological valence corresponds to the number of grammes of body protein which can be replaced by 100 grammes of the food protein in question. The evaluation of the "limiting amino-acids," which were most poorly represented in the protein in question, was carried out in this way. According to this method of approach animal proteins generally have a higher valence than vegetable proteins. The basis of this theory could now be shaken.

protein, instead compounds of two proteins always have a higher valence than their component parts.

The earlier assumption that good nutrition depended entirely on high protein content, and that foodstuffs which contain small amounts of protein like potatoes (2 per cent), are practically useless in terms of protein content compared with meat (19–21 per cent), has now been disproved. In fact the small protein content of potatoes can contribute so much to the increase in valence when combined with other foodstuffs, that the overall protein requirement falls surprisingly low.

It should be recognized that academic medicine, on the other hand, still considers a high protein minimum (1 gm/kg body weight per day) to be absolutely essential. Recent findings now give support to the pronouncements of dietary reformers, who have been advocating for years that we should give up meat and reduce our protein intake. Kofrany did not even exclude the possibility that with a purely vegetable diet the protein requirement could be lowered well below 24·7 gm (0·75 ounces) a day. Dr Ralph Bircher has therefore been able to speak of "a true revolution in dietetics" in academic medicine (17).

A vast number of further theories have developed out of Dr Bircher-Benner's teachings. I shall only mention a couple of them here: Are Waerland (139) (1876–1954) advocated a purely vegetable diet of cereal and milk products. Meat, fish, eggs and all luxury foods should be given up altogether. Like Dr Bircher's, Are Waerland's work is also embodied in a wide range of writings and publications. Dr Kollath (1892–1971) should also be mentioned for bringing a more scientific approach to nutrition (75). The Kollath breakfast consisted of whole flakes of wheat prepared in different ways with milk, yoghurt, grated apples, lemon juice and honey.

A whole range of other diets exists, all of which represent progress in one way or another. But somehow the totally organic viewpoint is often missing. The following interpretation is common in modern dietetics: a healthy diet is an indispensable prerequisite for health and efficiency. Illnesses are not entirely attributable to chance or fate but appear sooner or later as the result of mistakes in nutrition or life style. Of course, environmental factors are also crucial for health, so that in certain circumstances a single factor can have a very profound effect (for example, injury caused by radioactivity from atomic bomb tests or emissions from nuclear power stations). Moreover, our average life expectancy has already been partly predetermined by the life styles of our ancestors. However, this does not mean to say that we are doomed. On the contrary, every single person has the right to try to prolong his individual life expectancy for as long as possible by means of a suitable diet and way of life, and by compensating for environmental influences as far as he can. Diet can play a vital role in this.

Foods should not be judged solely on chemical or physical criteria; that is, on the well-known basic nutritional elements they contain: carbohydrates, proteins, fats and other essentials (vitamins, minerals, unsaturated fatty acids, trace elements, aromatics, auxomes). A knowledge of these substances is certainly helpful in understanding foods but their organic value—including freedom from poison—is much more important. The less a food is changed by man, the more successfully the naturally designed system of vegetable and animal nutrition is maintained—in other words, the basic elements, along with a number of other vital substances (some adequately researched, some not), must remain in a harmonious biological balance.

Those who have had no contact with the essential principles of natural healing may not be able to understand, or believe, Dr Fritz Becker, the head of a clinic in

Berchtesgaden, who maintains (12): "Fifteen years' work in the field of natural healing has brought me so far forward that today I am in a position to treat my patients without recourse to the chemical-pharmaceutical industry, purely with naturally functional ingredients. . . . I am proud of the fact that there are none of the usual medicine bottles in my establishment." But the patients who come to his clinic are usually the "least grateful" for, as Dr Becker writes, ". . . most people only decide to take the correct path to a healthy life, by radically altering their diets, when they have already tried everything else. Having gone about things the wrong way for so long, they have wasted not only most of their vitality but most of their bank balance as well."

Dr Becker sums up the most important rules for a healthy life very simply:

Regular intake of natural vegetables, cooked as little as possible and without artificial additives;
regular intake of pure fresh air and pure water;
regular physical exercise for all parts of the body;
positive thought and action.

Natural healing has intentionally been given quite a lot of space in this section because ignorance in this field is still far too great. A number of blunders by trained doctors in carrying out these techniques have also brought natural healing into some disrepute. But it can lead people—healthy as well as sick—automatically to a natural way of eating and, therefore, to a preventive approach to caring for life and health. In treating illness, optimum results can be achieved by a sensible compromise between natural healing and academic medicine.

Only by understanding natural healing can we really grasp the overwhelming importance of a natural diet, even for healthy people. Chemical and radioactive contamination of our food cannot be allowed to continue any longer. The whole world has now become aware of the dangers of chemical poisons. Everybody knows, for example, about lead, mercury and DDT. But, on the whole, little attention is yet being paid to the dangers of radioactive contamination. Even the radioactivity which resulted from the atomic bomb tests did not really penetrate the public consciousness, because it was hushed up or played down by official departments.

However, even though, fortunately, this radioactivity is decaying, further emissions of radioactive fission products into our environment are already taking place—from nuclear power stations—and those responsible for them are calculating that they will be diluted through natural processes, as has always been the case, although the question is really one of cumulative poisons.

What is the use of the Brugger Federal Council, speaking of the opening of the first Swiss nuclear power station, Beznau 1, saying that the Government had set up several commissions in order to fulfil its duty of protecting the state, amongst them a committee for the safety of atomic installations? In particular the Government would check whether all the safety measures were being carried out "according to the state of contemporary science and technology" (87). Really it all sounds most reassuring: until one notices the little loophole—"according to the state of contemporary science and technology." That means that there are still a lot of unknown factors involved. Such formulae lull the public into a false sense of security; and it is the public who will have to pay the price.

The history of science certainly contains grandiose successes to boast of, but it also includes a large measure of serious mistakes. When the first helicopter was spraying the first DDT on the fields, it was spraying it "according to the state of contemporary science and technology." Today the whole world is contaminated

27

with DDT. Similar contamination with radioactivity from nuclear power stations—with official sanctioning and in the face of numerous protests—is just around the corner: and then there will be no turning back.

As nutrition plays such an important part in human life people should be made aware, in the light of our present knowledge, of what can be expected from the contamination of food by radionuclides. Dr D. Kistner (Federal Research Unit on Food Preservation, Karlsruhe) writes in one of his works (74): "The dangers to health which are resulting or will result from an increased radioactive content in our food cannot be predicted today with perfect certainty."

In the nutrition field, tolerance doses are quite unreliable, because the accumulation of radioactive isotopes absorbed through food leads to extremely dangerous internal irradiation of the body. The cumulative effects depend on the nutritional complex in which the radioactive particles are absorbed. Other crucial factors are (82): the path the radioactive substance takes through the body, the place, and the duration of accumulation. In spite of a great deal of thorough research, the gaps in our knowledge are still enormous. In considering this problem it should never be forgotten that any dose of radioactivity is fundamentally hostile to life. Organic damage begins with a radiation dose of zero (131) (see also p. 64).

Moreover each individual has his own particular eating habits and his own way of life: so that any estimate of risk which is based on the standard human being, who eats standard food, is bound to be unsatisfactory. This kind of estimate takes no account of the individual risk. *Apart from this, there is a lack of adequate scientific evidence, from the results of purely physical and chemical measurements, from which to try to predict responsibly the effects of radioactivity on our health and on our genetic heritage.* This will be demonstrated clearly in what follows.

6 · WATER POLLUTION

Ninety-seven per cent of the earth's water is sea water, two per cent is ice and only one per cent is fresh water. This fresh water, which is essential to all life forms, we use as a dumping ground for our waste, in rivers and lakes. But the natural waterways constitute gigantic ecosystems; and water bacteria—decomposing agents— have been occupied for millions of years keeping this environment of water organisms healthy.

This self-purification process was entirely adequate until man began to overload it with waste products and to over-fertilize it. The disturbed ecosystems are now breaking down. The waterways are becoming increasingly polluted and will ultimately become incapable of offering the water organisms suitable living conditions. The waterways are threatening to "die."

The interaction of chemical, physical and biological factors brings about the self-purification of water. The purification is primarily attributable to biological factors, i.e. the activity of thousands of millions of microorganisms—tiny plants and animals. Sterile water can never purify itself.

Organic, putrescent filth serves as food for the water bacteria and is therefore partially transformed into bacterial mass. A bacterium has a diameter of 3–5 thousandths of a millimetre and one cubic centimetre of water can support several million bacteria. They multiply incredibly quickly: each bacterium divides once every 20 minutes. Theoretically 1 kilogramme of bacteria could become 400 tonnes in 24 hours.

The reason that we have not been overwhelmed by bacteria long ago is that certain factors exist which limit unrestricted reproduction. An irrevocable ecological law operates here which ensures that neither trophic levels (in this case the decomposers) nor populations can ever grow beyond a certain fixed size; for, otherwise, the whole ecosystem including the population in question would collapse. It is this law that man has evaded by his increase in population—so that the whole biosphere, man included, is threatened.

Putrescent human waste which is in the water, becomes bacterial material besides being transformed into carbonic acid and water. To some extent it also fertilizes water plants which, in turn, produce the oxygen necessary for water life by biogenic aeration. Of course, oxygen also reaches the water through diffusion and turbulence.

In the living communities of the water, as in all ecosystems, the constant law of "eat and be eaten" reigns supreme. The larger animals always eat the smaller ones. Bacteria serve as food for the protozoa, which in turn feed the rotifers and small crustacea, which eventually fall victim to insect larvae and fish. But even the protozoöns play their part in the self-purification of the water by promoting flocculation of the colloids and so clearing the water. Mussels and sponges sieve murky water and nematodes and chaetopodes filter the sludge.

Everything in this ecosystem is wonderfully organized. The well-balanced food chains of these living communities have provided them with security, and the possibility of further development through natural selection, since the beginning of time—of course, as everywhere in Nature, by means of a vigorous turnover of individual lives.

At the very end of the food chain comes man who gets back—in the fish he eats—his own waste products: the faeces and kitchen waste he empties into the drains along with bath water, washing-up water, et cetera. In sewage they are partly transformed into bacterial substances and the whole cycle begins again.

Man has imitated, in organic sewage farms, the natural self-purification process. However, he only uses the first phase of the process, namely the transformation of waste into bacterial substances. But, unlike the natural process, it is not converted and channelled back into the food chain, but is left in the form of sewage sludge. The purifying technique employed by man is therefore only an imitation of natural processes; furthermore, it has to take place in the smallest possible space, in the shortest possible time, under economic conditions and with the greatest possible efficiency.

All water organisms have developed individual sets of conditions for existence in their environment. In the course of millions of years they have become used to these conditions. Above all they require special water temperatures and a particular oxygen and acid content (pH) in the water. Trout, for example, require at least 7 mg/l of oxygen, as opposed to rays which are satisfied with less. Healthy river conditions demand at least 3–5 mg/l. The more highly developed the water organisms, the more oxygen they generally require, although each species has its own individual optimum level.

The more heavily sewage is overloaded with waste products, the more quickly the bacteria and lower forms of water life multiply. However, all the organisms need oxygen, with the result that demand soon outstrips supply in the water. The quality of the water is affected: and so, in their turn, are the higher forms of water life. Trout, for example, can disappear in this way.

The same end result can be achieved by raising the temperature of the water: because then the oxygen content in the water is reduced in the same way. The solubility of oxygen in water decreases with higher temperatures. So the same result can be brought about by overheating the water or by overloading it with sewage. Therefore we can speak of *thermal water pollution*.

Nuclear power stations pose problems of this nature, for two-thirds of the energy they produce has to be siphoned off as heat through special water- or air-cooling devices. Both of these systems have their disadvantages. Air-cooling, for example, with wet cooling towers, can lead to the formation of ice and mist in the vicinity of the power stations. Direct air cooling is also a possibility; but it entails having colossal superstructures.

The cooling process using fresh water from our rivers should be discontinued generally because it means that the quality of the water is being affected everywhere. Thermal pollution of water caused by this process has hardly been investigated at all; what evidence there is cannot automatically be presented as a concrete case, for each river has its individual character, with its own living communities and its particular chemical and physical conditions.

This is not to say that a certain amount of experimental work on thermal pollution by nuclear power stations should not be attempted. After careful calculation, experts have reached the conclusion that a limit of up to 3 degrees Centigrade imposed on water heating would be acceptable. The quality of the water, too, can be tested at any time by biological, chemical and physical analyses, so that we do have a reliable indicator of the organic effects of pollution at our disposal. Every river-cooled nuclear power station in operation should be seen as the beginning of a long-term experiment in thermal pollution —potentially a quite defensible situation.

However, this does not mean that experiments on a massive scale can be defended: because the consequences of radioactivity from nuclear power stations are incalculable. As we shall see later, radioactivity supposedly can be measured physically and understood analytically; but the biological consequences of such conclusions—however exact they may appear to be—still cannot be estimated satisfactorily. There is no indicator whatsoever for recognizing the onset of radiation damage. In addition genetic injury cannot be foreseen. As we have already said, the organic changes of fauna, flora and chemical action in the water can be constantly checked by river water experiments. The exchange of heat with the air can also be measured; so that, on the basis of all these results, heat loading limits can be established. The case of thermal pollution presents a complete contrast with that of radioactive contamination, for in the former, depending on the organic affects observed, science has the power at any time to say : "Thus far and no further!" If it appears that far-reaching damage could result it then becomes the business of the authorities, and of politicians, to force atomic power stations to lower their heat output, or to change over to a different cooling system, or to face closure. Fresh, running water could then be restored to a healthy condition.

A good example of how this can be done was provided by the Swiss Water Conservation Department in spring 1971. After it had been ascertained that the water in the Aar and Rhine rivers no longer conformed to the legal norms of grade 2[1], fresh-water cooling using the rivers was abolished for projected atomic power stations. This courageous decision by the Water Conservation Department proved that it would not allow itself to be influenced by economic and industrial pressure groups. Even more gratifying is the news that the Swiss Department of the Environment, created in 1971, has become incorporated with the Water Conservation Department. Switzer-

[1] Grade: internationally the earth's surface waters are divided into four grades. Grade 1 contains the purest water.

land included environmental conservation in the 1971 Constitution.

The problems of thermal water pollution and cooling towers are often over-publicized and exaggerated—in comparison with the much greater dangers of radioactive contamination. This is hardly surprising: because we have reached the position today where we experience water pollution directly in our everyday lives. Water unsafe to bathe in, bad drinking water, dirty rivers and lakes, the fishing crisis, and the economics of sewage farms are all obvious, concrete facts. Compared to these, radioactive damage from constant small doses often does not make itself felt for years, decades or generations. It is just this that makes such emissions of radioactivity so insidious and irresponsible.

The world-wide situation with regard to water pollution looks bad, although in certain fields great progress has been made. Professor Wuhrmann (144), a biologist at the Eidgenössische Technische Hochschule in Zürich, predicts that the organic remains (left over after the purification process) which flow out of the sewage farms will lead to just as much water pollution in the 1980s as there was in the 1950s, when water purification had only just begun. On the premise that the present growth rates for population and industry remain the same, all this would still happen even if all the drains were connected to sewage farms.

Our present twin-level sewage farms—even the very best technical and organic ones—are supposed to have a purification success rate of about 95 per cent on organic materials which are potential pollutants—although, in fact, only about 50–70 per cent of all organic pollutants is removed. We will be overtaken by events unless we can evolve new sewage purification techniques and think up new control measures. This is only another way of saying that mankind cannot possibly support a population explosion. A third sewage purification level (e.g. chemical precipitation, absorption with active carbons and so on) will probably have to be added. Unwise "doctoring" of the countryside—clearing forests and diverting rivers,

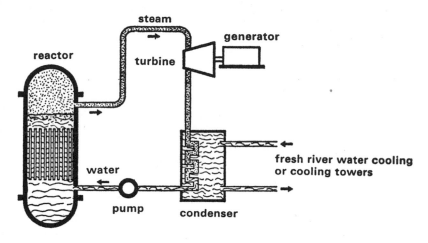

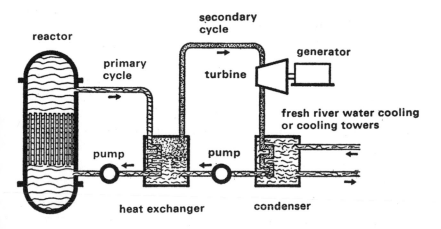

Fig. 2. Light water reactors. *Above:* boiling water reactor. *Below:* pressurized water reactor.

using measures which contribute extensively to the sinking of sub-soils—also lead ultimately to erosion of the land and to a deterioration in man's water resources.

Almost all of the dirt and poison eventually end up in the sea, which is now in the process of dying. When Thor Heyerdahl, crossing the Atlantic on a papyrus raft, noted that he had been sailing along for days on a sea of oil that made bathing impossible, he was only observing the superficial alarm signals. Had he dived down he would have noticed that there was no longer any clean water even at greater depths.

The French oceanographer, Jacques Cousteau, in a lecture to the European Council, made quite alarming statements about the pollution of the sea. The evidence for this is world-wide. Cousteau noticed on his diving expeditions that in all the seas the great coral reefs were perishing. Coral probably reacts particularly sensitively to changes in environmental conditions. Cousteau estimates that 30–40 per cent of marine life, especially plankton, has been destroyed in the last twenty years. In spite of this the sea is still being used as a refuse dump— even for atomic waste!

Today science knows that the self-purification processes of the sea are already overstrained, and that the seas are so polluted with materials that are hazardous to health that marine animals as far away as the polar regions have accumulated poisons in their bodies. It would therefore appear to be rather dangerous to use such animals for food.

However, a large section of the public and the authorities still lives under the delusion that the sea is inexhaustibly large. Yet by destroying the sea man is also destroying the biosphere, the sum of all the ecosystems on earth: on the preservation of which man as a species depends. He is therefore ultimately jeopardizing his own survival and his existence.

7 · AIR POLLUTION

Unquestionably changes in the atmosphere are also a part of environmental pollution. Rapid industrialization, the population explosion and the fantastic increase in the numbers of motor vehicles are, if we disregard radio-active contamination for the moment, the chief causes of the deterioration of the atmosphere.

In Switzerland, the most important source of air pollution is the motor car. Then come domestic fuels and, only in third place, factory chimneys and industrial plants: which are nevertheless, often the most dangerous source of pollution for their immediate surroundings.

Two main types of air pollution can be distinguished: pollution in the form of dust and pollution in the form of gas. Every imaginable thing swirls about in the air we breathe: sulphur dioxide, nitric oxides, aldehydes, volatile hydrocarbons (sometimes cancerous like benzpyrene), heavy metals (lead, mercury). The effects of these impurities can be manifold (113): increases in mortality rates through respiratory diseases, circulatory disorders, lung cancer, a higher incidence of bronchitis, inflammation of the lungs, asthma, and general listlessness and loss of vitality. On the present evidence it seems likely, too, that occasional genetic damage may occur.

Today every citizen of Zürich breathes in with the city air almost as much cancer-producing benzpyrene in one day as a smoker who inhales five cigarettes a day—in New York it would be as many as forty cigarettes. It will not be long before we see the inhabitants of some cities running about in gas masks.

The fine dust in the atmosphere makes a haze which filters the sun's rays. On the whole the meteorological effects of air pollution are still uncertain. The combustion of our fossil fuels (coal, oil and natural gas) increases the carbon dioxide content of the atmosphere, which influences the earth's temperature. Theoretically, because of the increased heat absorption, a temperature rise could be expected on the earth's surface which would melt the ice in the polar regions. On the other hand, a drop in temperature is also possible as a result of the haze effect which filters off the sun's rays. That would produce a new ice age. Experts are still uncertain, and are unable to predict, which factor will ultimately tip the balance, or to say whether the eventual consequences can ever be counteracted (113).

Burning fossil fuels causes a greater consumption of oxygen, and many people are already afraid that one day there will not be enough air left for our descendants to breathe. There is certainly some oxygen deficiency in heavily industrialized areas, which will have to be balanced by outside air flows.

At the present time it is estimated that 40–50 per cent of oxygen production comes from vegetable plankton in the seas of the world. Like all green plants, plankton converts carbon dioxide into sugar and starch by means of light energy (photosynthesis) and simultaneously gives off oxygen. Plankton levels have already fallen by many per cent as a result of marine pollution, so the production of oxygen will also decrease. DDT is believed to have had particularly pernicious effects in this respect.

The rest of the oxygen is produced by the green areas of the earth; but forests, bushes, hedges and fields are becoming scarcer and scarcer. In Switzerland, for example, one square yard of land disappears every 2–3 seconds because the ever-rising standard of living constantly demands more and more room for concrete, streets, housing estates, industrial installations, et cetera (145).

However, the most worrying of all these unpleasant phenomena is not actually the reduced production of oxygen but the progressive exhaustion, or complete destruction, of existing ecosystems on land and sea. By

38

constantly reducing the green belts, we are draining away the life blood of many of the organisms that share our ecosystem. We are weakening our links with nature at the same time. Professor Dieter Hoegger, President of the Federal Commission for Air Purity, has presented the case for not over-reacting to the lack of oxygen (46): "In the plant-animal-man cycle, exactly the same amount of oxygen is consumed as is produced by the plants. In the natural processes of evolution and decay, oxygen is used within a closed circle. This thesis starts from the assumption that the whole of the vegetable substance, whether consumed as food, allowed to decay, or burnt, is totally decomposed. Whether or not this is actually the case we must leave open for the moment. On the other hand we know that, in earlier geological epochs, this process of decomposition remained partially incomplete. Material from dead plants and animals was not totally oxidized; the oxygen produced by the plants in question during their life span was not entirely consumed. Instead, substances were produced which we today term fossil fuels; at least some of the free oxygen in the atmosphere was also produced in this way. The oxidation process, which remained incomplete in the evolution and decay of nature, can be completed today. In this way the fuel and the oxygen (which have remained separate for millions of years because of the interruption which occurred in the natural processes) can now be used. The vast reservoir of atmospheric oxygen, which remains outside the present cycle, is being used today."

The present oxygen content of the atmosphere amounts to about 21 per cent. According to Hoegger, it will take 8500 years for the oxygen supply to sink by 10 per cent, i.e. from 21 to 19 per cent. However this estimate is based only on the consumption of the oxygen that has remained untouched for millions of years because of the incomplete decay of fossil fuels. It is significant that

39

the existing oxygen deficit is often over-dramatized: for it represents an ideal argument for the justification of nuclear power stations, which use no oxygen.

8 · CUMULATIVE WASTE

Every healthy ecosystem has a natural biological cycle, in which the decomposing agents recycle waste material. In this way, waste material can be used to build new life in a very short time.

But mankind today has flouted this ecological law, and overpopulation is placing too great a strain on the decomposing agents. Moreover, chemical poisons damage both the decomposing agents and the food chain. Many of the waste products from technological and industrial plants cannot be disposed of in a natural way, either because they are biologically difficult to break down or because they cannot be decomposed at all. New synthetic materials like polyethylene and PVC (Polyvinyl chloride) are particularly resistant to all micro-organisms and it is estimated that it will take millions of years to break these products down.

A "waste explosion" now threatens all the major cities of the world. The American Health Service puts the amount of solid waste deposited in the USA from animal, inorganic, agricultural and industrial refuse, as well as from private homes and state institutions at something like 100 pounds per person per day. The situation has recently been exacerbated by our ever-expanding "throw-away culture" because many used bags, containers and "disposable" bottles are made of synthetic materials or aluminium.

Because of the way our civilization has developed, human beings can no longer rely on the natural biological cycle to provide the ecological conditions necessary for life. Refuse should no longer be thrown into rivers and lakes or tipped into the seas. Instead man will have to create an artificial, technological process to co-exist with the natural, the recycling process.

There are three basic processes involved in the initial treatment of solid waste: controlled dumping, incineration and organic decomposition. If possible the process must fulfill two crucial tasks: namely, remove the waste and regenerate the raw materials so that they can be used again. Of course, we must accept that losses are inevitable in the cycle of every raw material, so that in the very long term the disposal of industrial and technological waste cannot represent a genuine cycle. It is in this context that we should view the current massive despoilation of land at the expense of future generations, in spite of the fact that we have hardly anything in the way of an industrial regeneration process for raw materials. Many rare metals, which are crucial for our technology, are often lost forever in this way. The three processes mentioned above—dumping, incineration and organic decomposition—are, in principle, the methods which should solve the present waste problem. Dumping is not the ideal solution, for it does not actually get rid of the waste but simply gets it out of sight. Waste is often dumped in geologically adapted subsoils in artificial or natural basins where it is sandwiched in layers of soil and finally covered over with grass. This process avoids vermin and smells but the raw materials are lost forever. Moreover the possibility of underground pollution of water from the dump cannot be ruled out.

However, the incineration of waste is also a compromise which does not provide us with the very best method of waste disposal. Waste is supposedly reduced by some 60 per cent by this method, but what is left remains as slag that nobody wants; and the raw materials have again been eliminated from the cycle. The production of heat energy could be considered a useful by-product of this method; but it only becomes a viable proposition with very large installations.

Unfortunately the proportion of synthetic materials in waste is constantly increasing; according to Professor R. Braun, ETH Zürich, it could amount to as much as 20–30 pounds per person per year in the future (18). 12·5 per cent of this can be put down to PVC. When burnt this produces hydrochloric acid, which escapes into the air, although it is partially neutralized by the alkaline substances in the waste. According to observations and measurements made at the ETH, there is no need to fear damage from the release of hydrochloric acid; according to Professor Braun, this problem should not be over-dramatized because it might delay or prevent the development of essential incineration installations for waste. On the other hand, the danger should not be played down; for the potential risk of air pollution and the resultant harmful effects on life from burning PVC cannot be ignored. The incineration plants themselves are subject to erosion from the formation of hydrochloric acid. It is vital that ways and means be found to neutralize emissions from incineration plants as far as possible.

In future industry must be compelled by law to make the solution of problems of waste disposal and regeneration an integral part of any new development schemes. Products should no longer be developed if their production implies insoluble waste-disposal problems.

Man has found a waste-disposal process to break down refuse which closely resembles the natural organic cycle. Domestic waste (preferably mixed with sludge), after metals and plastics have been removed, is subjected to a process of decomposition so that in a few months a kind of humus is produced which can be re-used for cultivating land because it makes an excellent fertilizer. Using sludge is also, in principle, an ideal technical-biological process. Sludge can be converted into a first-class fertilizer through putrefaction, so that, particularly after pasteurization, it can be used again on the fields. However, not all of these advantageous biological processes can be put into operation today on a world-wide scale. Their use depends on the type and proportion of industrial refuse in waste and sludge: because industrial refuse may prove difficult or impossible to break down, or it may be poisonous. In such a case its re-entry into the natural cycle could affect the taste of agricultural produce—quite apart from the fact that under certain circumstances food contaminated by chemical traces could be damaging to health.

The great danger exists that, with an increase in industrial waste, it will become more and more difficult to achieve the necessary rechannelling of waste products back into the biological cycle.

But man is already committing what could be his greatest folly, before these problems have been even partially solved. Instead of drawing the inevitable conclusions from the contamination of our biosphere and food chain by the radioactivity from atomic bomb tests, he is continuing to use nuclear fission and releasing artificial radiation from nuclear power stations into our environment—and, consequently, into the biological cycle.

The layman is lulled into a false sense of security by those who talk of official control, of strict supervisory measures with maximum threshold doses, et cetera. In the long term all these measures are ineffective, for a number of reasons: first, organic damage can be caused by a radiation dose zero; then, many artificial radionuclides accumulate in the biomass; finally, nuclear power stations are being built in ever-increasing numbers with ever-greater outputs. Risk is further increased in the event of a break-down or catastrophic accident, when vast amounts of radioactive substances could be released.

These pernicious materials remain in the natural cycle for decades and centuries, and would lead in the end to a never-ending process of self-poisoning.

9 · ATOMIC WASTE

"Extreme safety consciousness from the word go" is used as a catch phrase to make atomic power stations sound acceptable to the general public. Certainly a great deal of trouble has been taken to try to eradicate sources of danger as far as possible; and, in part, this has been achieved without too much expenditure of effort or money. However, this is unfortunately not enough to put an end to radioactive emissions.

A further problem which must have appeared insoluble almost from the beginning is that of atomic waste. In spite of this, atomic projects have been carried out recklessly for the last twenty years. Even today nobody knows what will happen to atomic waste in the long run. The "extreme safety consciousness" does not seem to have amounted to much in this connection!

The fact is that more massive amounts of radioactivity are produced inside a reactor the longer it is in operation. Some conception of the size of the potential risk can be gathered from the following comparison: a nuclear power station of 350 megawatts, which has been in operation for a considerable period, will have accumulated a quantity of radioactivity amounting to about one thousand million curies (142). Quite apart from the elemental make-up and the long life of the atoms this corresponds to an activity of about two million pounds of radium. The total amount of radium produced in the whole world so far can be measured in pounds. Try to imagine the colossal dangers which threaten us now—in particular, remember that all this waste will have to be isolated from the biosphere.

44

In fact radioactivity cannot be artificially destroyed, in contrast with other well-known poisons which can be burnt or rendered harmless by chemical methods. Radioactivity decays at a rate determined by its own law of half-life. The half-life of radioactivity is the time taken for the activity of a radioactive substance to decay to half its original strength: that is, for half the atoms in it to disintegrate. This activity—depending on the type of atom involved—can take seconds, years, centuries or thousands of millions of years.

Every possible method has been tried to isolate atomic waste, even methods which were totally unsuitable from the start. The oldest method consists of dumping the waste in the sea in steel containers. Although detailed regulations have been laid down for this, it is quite clear that nothing can prevent the containers from corroding or guarantee that they will not combine with the waste. Even if waste is embedded in glass or cement containers first, absolute safety cannot be guaranteed. Dumping in the sea is still carried out today, in Great Britain and Japan, for example. In the summer of 1971 about 2600 tons of European atomic waste were dumped into the Atlantic about 420 miles from the coast of Europe (7). Thousands of tons had already been dumped in the same place.

At the information session on the safety of nuclear power stations held by the Swiss Atomic Energy Commission in Berne, on the 4–6 November 1970, dumping in the sea was defended, and upheld as a practical method of waste disposal (76). In opposition to this Professor Jaag, ETH Zürich, condemned the process very strongly. He said (73): "It seems to me absolutely indefensible that the authorities should sanction the dumping of atomic waste and highly toxic substances on the sea bed, in containers which cannot be guaranteed in the long term to remain impervious, so that sooner or later the poisoning of sea water and the resultant accumulation of toxins in marine life will become inevitable."

Radioactive waste is also disposed of by burying it. Although strict regulations exist for this and the ground is carefully selected it cannot be overlooked that this process does not offer complete safety either. So other potential solutions are being sought.

Very high-activity waste in liquid form is produced by the so-called "reprocessing" of used reactor fuel elements. The liquid is boiled down and stored in tanks. Soon more and more of these tanks accumulate, containing millions of gallons of liquid. Constant supervision and cooling are required if the contents of the tanks are not to boil over. Moreover, a percentage of the tanks have to remain empty to that the liquid can immediately be re-stored in the event of any of the tanks becoming pervious, which happens repeatedly. Dr Schultz notes (100): "It is curious to think that if highly active waste of this kind had been dumped some time in the fourteenth century (about the time of Marco Polo, Dante and Columbus), protective containers and constant supervision would still be necessary today." So the next generation is doomed to a sinister inheritance. Just think of having to supervise waste left by Napoleon! One part of an American waste installation has been sited in an area threatened by earthquakes, and efforts are now having to be made to stabilize this liquid waste.

Burying the waste in stoney ground and concrete dugouts was not a great success either; neither was sinking the waste in disused mine shafts, for experience showed that the radioactivity always managed to find a way back into the environment. In the USA and in Germany efforts are being made to deposit radioactive waste in deserted salt mines or rock salt formations. However, experiments are not scheduled to begin for some years; and what use are tests that last for a few

46

years, when disposing of the waste is a process that lasts for centuries? Nobody can predict with any certainty what sort of effects concentrated radioactivity will have on the rock salt formations over a long period of time. Apart from this, the radioactive material could come into contact with subterranean waters, or tectonic changes might take place.

The medical associations of the Swiss cantons of Baselland, Baselstadt and Aargau expressed their position on the question of atomic power stations as follows (27): "Before nuclear power stations become operational, definitive regulations must be established for the safe disposal of the used radioactive fuel elements. Planning based on vague suppositions is totally inadequate." But it is in this particular field—the disposal of atomic waste— that ultimately we have to rely most heavily on supposition! These doctors are nevertheless to be congratulated for their courageous stand. If it is objected that Switzerland has no atomic waste problem in terms of highly active substances, this is just an extreme example of dodging the issue. In fact the used fuel elements are simply sent abroad to be "reprocessed." In this process only certain useful materials are removed and retained, leaving the rest—still an insoluble waste-disposal problem. So Swiss atomic waste is not finally disposed of—the problem is simply moved out of Switzerland and sent abroad.

The problem of atomic waste was discussed with a similar lack of responsibility at an information session of the Austrian atomic conference held in Vienna in March 1971. There it was stated that (40): "The disposal of radioactive waste is not a terribly serious problem for Austria because the overwhelming majority of used fuel elements are exported. Their ultimate disposal is undertaken by the countries which reprocess the fuel elements."

So the problem of atomic waste really represents one of the saddest chapters in the history of the peaceful industrial application of nuclear fission: we must now face the fact that nuclear industry has been producing indestructible waste of the most deadly kind for the last twenty years, and the problem of its long-term control still remains unsolved.

At the beginning of 1972 the new director of the American Atomic Energy Commission (USAEC), J. Schlesinger, is supposed to have put forward the theory that in ten years' time we would be "shooting" atomic waste off to the sun in "spaceships". What might happen if the "spaceships" broke down does not bear thinking about: all the waste could come hurtling back to earth as fallout!

II · Atomic power stations and their dangers

As the previous chapters have shown, progress—by which is always meant material and technological progress—has led us up a blind alley, from which some way out has to be found if man is to survive. As far as many of the new technological processes are concerned economizing has been the rule, and, as a result, waste products have simply been dumped on our environment. For a while the environment was able to dilute or neutralize this waste; but the task soon began to overtax natural resources. The consequences of this—pollution of land, air and water, and the poisoning of our food—are becoming more and more serious. If we carry on in this way it is obvious that mankind will commit collective suicide in all its technological splendour. Fortunately, however, people are becoming increasingly aware of the situation. The question of preserving life and the environment has come in for more discussion in the last few years and is increasingly being taken into account by legislation.

Now, however, man has begun to meddle with the radiological balance of his environment, before the damage he has already inflicted on nature has even partially been put right.

The dangers posed by the current light-water reactors are quite unprecedented:

Production of radioactivity on a massive scale, the biological effects of which have not been properly estimated;

the risk of a catastrophic accident cannot be ruled out entirely; if safety measures should fail the calamitous effects of this could be much more disastrous than people now realize;

the emission of radioactive substances into our environment, not just through major or minor accidents (which cannot be prevented with any degree of certainty) but also from the normal operation of the reactor;

the production of indestructible radioactive atomic waste and effluent;

thermal water pollution;

climatic changes caused by cooling towers (these last two effects occur with all power stations except hydroelectric).

Radioactivity can cause diseases and complaints that are very difficult, or even impossible to cure like cancer, leukaemia, tumours of the bone marrow, premature ageing, lowered resistance to infection, reduced vitality, diminished fertility, physical and mental damage to the embryo, et cetera. There is no prophylactic treatment against these ills. The most insidious effects, however, which may not become apparent for years or even decades, are those caused by continual small doses of radiation. Damage to the human genetic structure is concealed, and does not emerge until generations later; at the present time there is no known way of tracing these effects (137). In the case of most diseases radiation damage begins with a dose rising from zero (66, 131) and the establishment of tolerance doses—or as they are generally known today "maximum permissible doses"

(MPD)—is only intended to restrict the increase in disease (as caused by human interference).

Generally it can be said that the precise scientific data needed to calculate radiation levels are not even available. (55). Nevertheless the nuclear energy industry entrenches itself behind the argument that it is maintaining the prescribed limits, or even congratulates itself on keeping below them.

Like all the other evils of civilization, nuclear technology could ultimately compromise between risk and profit. However, from the radiobiological point of view, so many of the crucial elements needed to calculate the risks involved are shrouded in mystery that the search for a compromise simply cannot remain in scientific hands, but must be extended to the realms of politics, philosophy and individual conscience (49, 50, 53, 137). When the fate of whole population groups and future generations is at stake no compromise can possibly be reached if a number of factors remain unknown. Our generation must not strive only to find a transitory solution to the energy problem at the expense of future generations. In any case, our own generation is already suffering from it.

Until now only a relatively limited number of people have concerned themselves with the dangers arising from the utilization of radiation. Most of them have been people working in science and medicine who are concerned with the industrial applications of radiation, like the use of radiotherapy for the diagnosis or treatment of illness. Naturally, vital examinations and treatment should not be stopped simply because of anxiety about the dangers of irradiation: but the medical world has become more cautious, and the modern doctor has to take these dangers into account. The patient will often accept the risk of irradiation as a hazard necessary to his recovery; but nobody can compel him to take that risk.

51

The present legislation on radioactivity does provide a valuable guide for the industrial applications which we have mentioned, although it rests on inadequate scientific foundations. *The theory of critical organs on which the present laws of the International Commission for Radiological Protection are based has recently been proved to be largely inaccurate and quite unsuitable as a basis for the calculation of risk* (67).

With the development of nuclear energy an entirely new situation has arisen. Artificial radioactive substances are now being released into our environment and nobody can escape their effects. Apart from this, the radioactivity that remains in the reactor represents a potential menace which did not exist before, for technological or human break-downs, malicious damage from war or sabotage and natural disasters can never be completely ruled out, even with the most refined safety measures (90).

A situation now exists in which every individual and his descendants will be forced to face risks from artificial radioactivity which cannot even be adequately calculated. The radiological protection laws cannot do their job in the face of the existing global dangers.

1 · PROMINENT PEOPLE WHO HAVE SPOKEN OUT AGAINST NUCLEAR POWER STATIONS

Deliberately, the arguments in this book are not based mainly on the statements of prominent individual scientists. The supporters of atomic power are forever objecting that their opponents are only outsiders, laymen, health fanatics or people wanting to stem the tide of progress, and even the possession of the Nobel prize does not seem to offer much protection against such accusations. We will be making use here of the very sources upon which the atomic industry itself rests and must rest, namely the International Commission for

Radiological Protection (ICRP) and the working groups set up by this commission. The radiological protection laws of almost every country are based on the recommendations of this international commission. A further unimpeachable source of information can be found in the publications of the United Nations Scientific Commission on the Effects of Atomic Radiation (UNSCEAR) in their *Reports*. This committee supplies in abundance the best proof of just how underdeveloped radiobiology still is today, with the result that a serious estimate of the risk to mankind from additional amounts of radioactivity is simply not possible.

First, however, in addition to considering the evidence of learned bodies, reference should be made to a number of prominent scientists, doctors and other personalities, who have taken up a contrary position or who have warned against the dangers of nuclear power stations.

Above all, the Gesamte Östereichische Ärztekammer (Joint Austrian Medical Council) should be mentioned; in an exhaustive memorandum it took a stand against the building of the first nuclear power station in Austria. Austrian doctors are to be congratulated for this courageous act. But why are all the other medical associations silent? Surely they are responsible for public health too?

Here are some statements by opponents or critics of nuclear power stations:

Professor LINUS PAULING, Nobel Prize Winner for Chemistry (23): "There can be no doubt that the radioactive material released by nuclear power stations is injurious to the human race and will cause increases in the numbers of children born with severe physical and mental damage."

Professor ROBERT ROBINSON, Nobel Prize Winner for Chemistry (23): "The English nuclear power station at Calder Hall works uneconomically. The results only appear so favourable for the reason that a very high price is paid for the plutonium produced. I am not in favour of producing electrical energy in such a crack-pot fashion."[1]

Professor BENTLY GLASS, a biologist at the Johns-Hopkins University (23): "The so-called peaceful application of nuclear energy will cause mankind even greater harm than the military application has caused up until the present time."

DAVID E. LILIENTHAL, former first president of the USAEC (American Atomic Energy Commission) says (23) it would be irresponsible to build reactors in urban areas. He would not like to live in a king's palace if a reactor were to be built on the doorstep. From being an enthusiastic supporter of nuclear fission, Lilienthal has developed into a relentless opponent.

Professor BORIS RAJEWSKI, Director of the Max Planck Institute for Biophysics takes the following view (23): "We are not in a position to judge the effects of irradiation or even to estimate them. Looked at strictly scientifically, everything that has been published about the subject can only be considered as supposition or fantasy, or even credulous logical inference, all of which, however, are based on quite irrelevant premises."

HAROLD PRICE, Member of the USAEC, said in Congress in 1967 that, in general, actual experience with large reactors was still very limited. It would be wise to build them in places where protection could be offered by distance.

Professor WALTER HERBST, a radiobiologist at the University of Freiburg in Breisgau and well known for his

[1] The statement about Calder Hall is true, but Calder Hall was always intended primarily as a plutonium factory (for atomic bombs). But because there would inevitably be a large amount of heat to use up, it was decided to make it into a combination of plutonium factory and electrical power station. In the second capacity, it was inevitably uneconomic, but provided an enormous amount of information for later nuclear power stations (Ed.).

54

clear-headed scientific articles and lectures, also condemns the irradiation of foodstuffs (45).

Also in the ranks of opponents are: MAX BORN, Nobel Prize Winner for Physics; Professor KARL BECHERT, atomic physicist, Member of the German Parliament and from 1962 until 1965 the chairman of a government atomic energy commission; Professor WALTER HEITLER, one of the directors of the Institute of Theoretical Physics at the University of Zurich, winner of the Max Planck medal, winner of the Marcel Benoist Prize 1970; university professor Dr H. THIRRING, Vienna; Professor MAX THURKAUF, University of Basel. Heitler, Thirring and Thurkauf are co-signatories of the Austrian doctors' memorandum. Thurkauf is also well known for numerous articles against nuclear power stations.

Various associations have been drawing our attention for years to dangers like environmental contamination and radioactivity which are inherent in our civilization. Reference is made here to: Weltbund Zum Schutze des Lebens, with branches in 75 countries, which was founded by Professor GUNTER SCHWAB in 1958. This association came to prominence with a very comprehensive series of articles. There is also the International Society for Research on Nutrition and Vital Substances, which has a membership of some 400 scientists in 91 countries, among them six Nobel Prize Winners. The president is university professor Dr ADALBERT SCHWEIGART, Hanover-Kirchrode. Schweigart is also president of the international branch of the Weltbund Zum Schutze des Lebens. These groups have undertaken pioneer work.

Professor Schweigart can say, with justice (103): "The issue of the contamination of foodstuffs which we raised has been minimized and played down in exactly the same way as our protests against radioactive contamination of

c

the atmosphere; against the poisoning of the air with the exhaust fumes from industry and motor vehicles; against the saturation of drinking water with undesirable chemical substances; and against the spread of pesticides all over the world. Even the existence of diseases brought about by civilization was denied, until now, fifteen years later, a general 'Twilight of the Gods' seems to have descended." In the general resolution, which was prepared by the scientific council of the society in the presence of scholars from 78 eastern and western countries, and finally constituted at the fifteenth Vital Substances Convention in Luxembourg, this was said on the subject of the rapidly deteriorating situation with regard to toxic substances: "A great number of reactors are still being built in densely populated areas without adequate guarantees for the safety of the population, which no doubt results from the fact that far too little attention is being paid to the potentially horrific consequences which confront us."

We should also mention public health associations in many countries, although today they are still mocked and scorned by those who do not understand the real situation. Now, in the light of modern developments these organizations have come into their own. They are decided and decisive opponents of nuclear power stations.

Unfortunately many worthy associations and societies for nature conservation have concerned themselves very little in the past with the dangers of nuclear power stations. As nature conservation has obviously become a question of the conservation of life, too, one can only hope that this situation will soon change. A good beginning seems to have been made with Ernst Zimmerli's book *Tragt Sorge zur Natur*, which was jointly financed by the government and the Nature Conservation Society of Aargau (Switzerland). This highly readable book was published especially for teachers and children, as a con-

56

tribution to the European Nature Conservation Year in 1970. Talking about the dangers of the use of nuclear energy for peaceful purposes, it concludes with this: "How questionable here remains the use of threshold doses. When will man see that, with each step he takes forward that disturbs the natural equilibrium, he is undermining the foundations of his very existence?"

2 · THE RUDIMENTS OF NUCLEAR PHYSICS

(a) Atomic structure

A simple outline of atomic structure is necessary to any consideration of the dangers of nuclear power stations. In the discussion which follows this will be done in a way that the layman can easily understand. Detailed knowledge is not actually crucial to the formation of a sound and objective judgement, although people often insist it is.

Try to imagine an atom as a very tiny particle consisting of a minute nucleus surrounded by electrons—negatively charged particles. The structure of an atom can be compared to that of the solar system with the sun (the nucleus) at the centre and the planets (electrons) orbiting round it.

The atomic nucleus is in the centre (the sun) and it consists of positive and neutral particles: protons and neutrons. These particles are held together by nuclear force. The number of positive protons in the nucleus generally corresponds to the number of negative orbiting electrons, so that the atom is electrically neutral. However, if there are more or fewer electrons contained in the electron shell around the nucleus than there are protons in the nucleus itself, then such an atom is said to be *ionized*, or an *ion*.

Almost the whole mass of the atom is concentrated in

the nucleus. For example, the mass of a hydrogen atom is almost two thousand times greater than the mass of an electron which is orbiting the nucleus. Also the atoms are practically "empty" for, if we can imagine a nucleus about the size of a hazelnut (about 1 cm), then the electron would be revolving round the nucleus at a distance of about 545 yards! The nucleus itself has a quite unimaginable density. 1 ccm of pure atomic nuclear mass would weigh the colossal amount of 240 million tons.

The number of protons in the nucleus of every chemical element in existence is precisely defined; it varies, from hydrogen—whose nucleus contains one proton—to uranium, for example, with 92 protons. On the other hand the number of neutrons in the nucleus can vary. But all these numbers are not particularly important. All one needs to know is: that *the atomic nucleus of each element consists of protons and neutrons. The number of protons always remains the same for a given chemical element, but the number of neutrons may vary. These related atoms are called isotopes.* So isotopes are chemically identical although they differ in mass number, i.e. in weight. In order to distinguish elements and isotopes the mass number is written after the element name; the mass number indicates how many neutrons plus protons the atomic nucleus contains. If we write "uranium 238" it means that the sum of protons and neutrons in the nucleus of this atom is 238, so the mass number is 238. In fact uranium contains 92 protons and 146 neutrons—238 particles in all. The isotope of uranium 238, uranium 235, actually contains 92 protons but only 143 neutrons, so it has 235 particles altogether and a mass number of 235. *The number after the name of an element indicates the number of particles in the nucleus.*

It is also important to note that an element and its

isotopes behave exactly the same way chemically, although physically they do not.

There are atoms called *radionuclides* whose atomic nucleus is not stable but disintegrates of its own accord without any external influence according to a law which is not yet understood. The disintegration of the atom—which gives off radiation—is called *radioactivity*. These nuclei are called *radioactive*. The unstable nucleus can transform itself according to exact physical laws.

Every radioactive element or isotope takes a clearly defined time to disintegrate. The time it takes to disintegrate is the so-called *half-life*, which is the measure used to indicate that one half of a radionuclide has disintegrated. The half-life can vary from split-seconds to thousands of millions of years.

We also talk about *natural background radiation* as it exists in nature without any human intervention. If radionuclides are produced by human intervention we then talk of *artificial radioactivity*. This means that it is possible to convert stable nuclei into radioactive ones.

The discovery of radioactivity was made first by the German physicist, Wilhelm Conrad Roentgen (1845–1923), at the end of the last century. The roentgen (or X–) rays used in medicine were named after him. They are produced by using electron tubes. Analogous rays are produced by the nuclear disintegration of radionuclides; they are called γ-rays (gamma rays). These are electromagnetic waves with the same physical characteristics as heat waves, light, ultra-violet and X-rays. They represent a pure surge of energy from the nucleus and are even more penetrating than X-rays because their energy is greater. In nuclear reactors concrete walls 6–10 feet thick are needed to provide adequate shielding from these rays.

Radiological research began immediately after the turn of the century with the development of radium and the work of Marie Curie. She was the first person to succeed in isolating 0.1g of the element. But she did not recognize at the time that the mysterious glow given off by this substance was connected with the emission of extremely dangerous rays. As a result she died of what was at that time an unexplained blood disease—leukaemia; she was one of the first victims of radioactivity with which man was just beginning to experiment.

(b) Types of rays

Research showed that when atoms disintegrated four types of rays could be distinguished:

1. α-rays (alpha rays). These consist of minute particles of the disintegrating nucleus containing two protons and two neutrons; this is equivalent to the nucleus of one helium atom. They are not very powerful and only penetrate a few centimetres into the atmosphere—and about 0·1 mm into living body tissue. They have a very strong ionizing effect in spite of this. This is very important if an alpha ray ever becomes lodged in body tissue because, in spite of its very shallow penetration, the radiating particle can damage the body cells severely by its intensive ionizing effect.

2. β-rays (beta rays) consist of electrons which come from the atomic nucleus. A neutron can disintegrate into a proton and an electron and the electron is radiated. Beta rays have a greater energy and can penetrate living body tissues to a depth of several centimetres.

3. γ-rays (gamma rays). As we have already noted, this type of radiation is not connected with particles but with electro-magnetic waves of great energy which differ from alpha and beta rays in that they can penetrate concrete, lead, and steel.

4. Neutron rays. Artificial changes in the nucleus, like atomic bomb explosions or nuclear fission in nuclear

power stations, produce an extremely dangerous type of radiation in the form of neutrons. This is tremendously strong and penetrating.

To sum up briefly: substances whose atoms disintegrate of their own accord are called radioactive. They emit α (alpha), β (beta), or γ (gamma) rays . Neutron radiation is produced when artificial nuclear changes are induced. Natural radioactive substances have existed since time immemorial—like radium and uranium. It is very important to distinguish between gamma rays on the one hand—a form of pure wave radiation—and, on the other, alpha and beta rays which are made up of atomic particles. Neutron radiation is also a radiation of particles. Because rays also have ionizing effects we talk of ionizing rays too.

(c) The measurement of radiation
Now that we have briefly described the nature of radioactive irradiation we must learn the criteria for measuring and judging its physical and biological effects.

1. *Activity*
The activity, or quantity, of rays is measured in *curies*. If the atoms of a substance disintegrate at a rate of 37 thousand million a second this corresponds to an activity of 1 curie. This is the rate for one gramme of radium. Because the unit of 1 curie is rather large for the measurement of most natural processes fractions of curies are often used for calculations, with the following abbreviations:

1 curie	=	1 Ci
1 thousandth of a curie	=	1 Millicurie (mCi)
1 millionth of a curie	=	1 Microcurie (μCi)
1 billionth of a curie	=	1 Picocurie (pCi).

The number of curies alone, without the precise

61

elemental content, tells us nothing about the effect of radiation, it merely provides a measure of the number of atoms disintegrating in one unit of time.

2. Radiation effects

The different kinds of radioactive rays can have very different effects. On the one hand they have an ionizing effect and on the other hand they can produce heat on contact with the body. *Ionizing effect*; ionized atoms behave chemically in a different way from neutral ones. Through radiation chemical reactions which do not occur normally can take place. These can damage living tissue.

The physical unit for measuring the effect of ionization is the *roentgen* (R). A radiation of one roentgen produces two thousand million ion pairs in one ccm of air at normal temperature and normal pressure. We shall not need to use this unit of measurement in what follows but it was introduced in order to give a complete picture of the technicalities involved in measuring radiation.

Heating effect = energy yield; radiation energy is transformed into heat on contact with any other body. The body then absorbs the kinetic energy. However, the term *radiation absorbed dose* indicates the amount of radiation absorbed by 1 gramme of the body in question, not the amount absorbed by the whole body. The physical unit for measuring radiation absorbed dose is called the rad (short for "radiation absorbed dose"). So *rad is a measure of the amount of energy absorbed by an animate or inanimate body from radiation.* One thousandth of a rad is the same as a millirad (mrad).

We should also note the variety of effects which radiation can produce on living body tissue : alpha rays have the effect of an intense bombardment from a short distance. Beta and gamma rays penetrate more deeply. The physical definition of a rad does not always describe exactly the effects that are produced on living tissue and

62

this is why another unit to measure effect has to be introduced. This is the *"biological effect dose"*, the *rem* (abbreviated from "roentgen equivalent man"). One thousandth of a rem is the same as a millirem (mrem).

The fact that a variety of biological effects are possible with the same physical effects (pure energy yield) can be understood more easily by studying the following analogy made by Manstein (82). If an area of living tissue is struck, first with a pointed object, and then again with a blunt instrument—with the same force—the results will be quite different wounds. Unless the sharp instrument has pierced an important aorta the wound will heal relatively quickly. But if the blunt instrument has caused crushing and bruising of the tissues the wound will heal more slowly.

Different kinds of radiation can also have different effects with the same physical dose (rad). In this connection we speak of "relative biological effectiveness."

1 rad of X-rays, beta-rays or gamma-rays is the equivalent of 1 rem; 1 rad of alpha rays — larger particles — (and heavily ionized = blunt instrument), is the equivalent of up to 20 rem.

Rem can be converted from rad by multiplying with a *quality factor* (QF):

$$rem = rad \times QF$$

For neutrons of unknown energy QF amounts to 10 (72).

This explanation should serve to demonstrate that all kinds of uncertainties exist in radiobiology, even in the definition of dose effect.

It is simply not possible to make a direct measurement of doses of radionuclides which have become incorporated in body tissue; extremely complex calculations and measurements have to be used. When in addition there is absolutely no uniformity in the radiation ab-

C*

sorbed dose—as is the case with the bones and with radioactive particles in the lungs, for example—the data become almost incomprehensible. On this point the ICRP says (47): "There are certain conditions of radiation exposure in protection work where the QF concept can only be applied with major qualifications. . . . The concept of dose equivalent for external radiation, when applied to bone-seeking radionuclides, presents a number of problems on which further research is needed."

We only want to point out here that the biological effectiveness dose of certain important kinds of irradiation can only be estimated very inaccurately. This really provides only the most primitive basis for making statements about harmful effects, for calculating risk, and for establishing tolerance doses.[1] Who has the right to lay down "limits" or "maximum permissible doses" for the whole population? *According to the ICRP biological damage begins at a radiation dose zero* (66): "where tumours and genetic effects are concerned it is generally accepted that no limit exists. . . . The effects are quantitative in character and depend on probability per unit dose as well as total dose over the whole range from zero up."

Because we have already accumulated dangerous radionuclides from the natural environment in our bodies it does not mean that we can just go on adding more. On the contrary, this fact provides the strongest argument against any further increase and so-called "maximum permitted doses" should not be permitted at all. Every discharge of artificial radionuclides into the air and in the effluent from nuclear power stations is clearly irresponsible and we should not be misled by the seemingly trivial amounts of the doses involved. They are cumulative

[1] The reader will notice that rad and rem often appear interchangeably. This is not really confusing if, for our purposes, 1 rad is generally taken to equal 1 rem.

poisons which can accumulate in our bodies through inhalation, through the skin and through our food.

(d) Artificial nuclear fission

If uranium 235 atoms are bombarded with neutrons the nucleus will split into two fragments on contact with a neutron. At the same time two neutrons are also set free from the original nucleus. If these two neutrons then come into contact with a uranium 235 nucleus, new fragments are produced and, among them, two more neutrons, so that there are then four neutrons. This process is self-perpetuating. More and more neutrons and contacts with atomic nuclei will result so that a chain reaction occurs. During this process of nuclear fission, · gamma rays of enormous energy are released.

In a short time this chain reaction can become explosive, as it does in atomic bombs. The colossal energy produced by nuclear fission is released in a split second. The fission process can be controlled, however, by mopping up the avalanche of neutrons with neutron-absorbing substances like boron or cadmium. This makes it easier to obtain a controlled yield of kinetic energy which can produce steam and electricity in the conventional way.

Atomic power stations today use uranium 235 as fissile material. Up to two hundred different radioactive *fission products* are produced from it. These are a few of them:

Tritium 3	12a
Krypton 85 (noble gas)	10a
Xenon 133 (noble gas)	5d
Xenon 135 (noble gas)	9h
Iodine 131	8d
Caesium 137	30a
Strontium 90	28a
Barium 140	12d

| Zirkonium 95 | 65d |
| Cerium 144 | 277d |

(a = years, d = days, h = hours; 12a, for example indicates a half-life of 12 years).

Generally, radionuclides with longer half-lives are the most dangerous, because a nuclear power station can never be leak proof: it constantly gives off "controlled" amounts of radioactivity a little at a time into our environment.

Liquid and gaseous discharges and waste originate mainly from the inevitable leaks (tiny holes and cracks) in the shells of the fuel rods through which the fission products reach the primary cooling system; from there, a proportion of the waste escapes through the ventilation systems, into the discharged air. The primary cooling system is of course constantly being cleaned by mixed bed exchangers but the used exchange substances themselves form atomic waste. [This applies to water-cooled reactors—as used in the USA—the process is different for British gas-cooled reactors. Ed.]

A 500 MW (megawatt) reactor can contain as many as 30,000 fuel rods. "The actual number of fuel rods with leaks in a particular reactor cannot be defined with any degree of certainty," says atomic expert Professor Tsivoglou in his report on the proposed boiling-water reactor at Kaiseraugst (Basel). The radioactive content of gaseous waste and effluent obviously depends on the leakage rate, for this determines the rate of escape in the exhaust chimney and therefore affects the radiation dose in the vicinity of the reactor. *It is easy to see from this that the discharge from nuclear power stations can vary a great deal.*

Tsivoglou (121) estimates a chimney discharge of 0·01 to 0·02 Curies second. But the upper limit of a heavy discharge of 0·05 Curies second is not unusual and would correspond to a discharge into the environment of radio-active substances of approximately 1·6 million Curies year.[1]

This is a good point at which to emphasize the fact that, in principle, exactly the same fission reactions take place in a nuclear power station and in an atomic bomb *must* explosion, (80). "Advocates" for nuclear power do not on *be* the whole take kindly to being reminded of such facts *nonsence* and, in reaction, accuse their opponents of over-reacting because they can still feel the shock of the atomic bomb in their bones! In nuclear power stations most of the products of fission do indeed remain behind in the fuel elements and in the decontamination plants but they then give rise to the unsolved problem of the ultimate disposal of atomic waste.

Additional radioactive *corrosion products* are constantly released from nuclear power stations. They can be traced back to the radioactivity which is created by neutron radiation in the structural materials of the reactor. This process produces the following isotopes:

Cobalt 60	5a
Manganese 54	314d
Iron 59	45d
Zinc 65	245d
Chromium 51	28d

All these corrosion products reach the primary cooling system where they are continuously removed by ion exchange along with fission products. However, the used exchange materials remain in the form of atomic waste. Because of leakages in the pumps, valves, ventilators, et cetera, and in the course of repair work, a certain amount

[1] At the American boiling water reactor in Monticello, it is estimated that a release rate of 0·05 Curies second would correspond to a radiation exposure cone of 25 mrem/year outside the operating area.

of this reaches the effluent. As complete purification is impossible in practice, or is never actually carried out, it is inevitable that waste products will be channelled into our waterways.

Then there are liquid and gaseous *activation products*, created by neutron radiation on impurities present in the cooling water. These impurities result from decomposition in the primary cooling water, air filtering in, et cetera. Activation products are mainly radio-isotopes like nitrogen, oxygen, and argon and tritium (121).

In spite of built-in systems for storing waste gas, (noble gases are supposed to be held back for a period of hours or days before being discharged so that their activity can be partially reduced), and activated charcoal filters, it is quite impossible to hold back *everything*. Noble gases in particular escape very easily because of their slowness to react.

Short-lived radionuclides can be most dangerous if they are present in high concentrations in the chimney discharge. Professor Tsivoglou (121) draws attention to the fact that—depending on the rate of leakage—the concentrations of Xenon 135 in the chimney discharge of the General Electric boiling-water reactors can amount to *10 to 27 thousand times the maximum permissible limit of radiation* for people living in the vicinity of a reactor! This occurs with normal waste control and processing systems in operation and in spite of a retention period of 30 minutes. The atomic experts simply assume that the waste will be diluted rapidly in the natural environment around the power station before it reaches the first human beings.

The Swiss atomic power station at Mühleberg (a boiling-water reactor) in the immediate vicinity of the city of Berne, operates with similar 30 minute retention periods and with an activated charcoal filter. But long retention periods and activated charcoal filters do not reduce the discharge to nothing. For this reason alone—and there are many others—nuclear power stations should be banned completely.

This heterogenous mixture of the most dangerous radionuclides is gradually being dispersed in our environment, although in deceptively small doses, and one day these deadly substances may end up on our plates via the food chain. When we absorb radionuclides with our food we hear nothing, see nothing and feel nothing. The gaseous products burden our air passages and our lung tissue. Under these circumstances the statement of atomic industry that noble gases are biologically inactive should not provide informed people with any consolation.

3 · NATURAL RADIOACTIVITY

Our whole environment and we human beings have been living from time immemorial in a "sea" of natural background radioactivity. This radiation cannot be avoided and has both internal and external actions.

External irradiation comes from two major sources: from outer space, i.e. from the sun or distant worlds through the atmosphere, or from the earth.

The part that penetrates the atmosphere consists mainly of gamma rays. However this cosmic component is weakened by the atmosphere so that its strength depends on its height above sea level. It is calculated that the radiation level doubles roughly every 1000 metres.

Radiation originating on earth (again mainly gamma radiation) comes from radiocative rocks and minerals. Active elements and their bi-products (potassium 40, uranium 238, lead 210, polonium 232, radon 222 and thoron), dating back geologically many millions of years, are largely responsible for this radiation. The radiation

69

can also vary considerably according to the composition of the ground.

The various natural radioactive substances in our environment also enter the food chain, and from there they penetrate our bodies: through inhalation and through the skin, as well as through food and drink. This radioactivity, according to its particular characteristics, can then be accumulated internally. It is not just the radionuclides in the ground that can be accumulated in the body so, too, can radiation from outer space which produces active hydrogen (called H3 or tritium) and carbon 14 through transformations in the outer atmosphere, as well as other secondary products. Moreover, active bi-products are also generated by the disintegration of radioactive noble gases and they too can reach the body, in conjunction with aerosols, in the air we breathe. We inhale the noble gases and they penetrate through the lungs and bronchia into the body, where they accumulate over a period of time. All these radionuclides are absorbed by human beings. The result of this is the total natural *internal irradiation* including not only gamma radiation but also alpha and beta radiation.

The effect of this natural background radiation on human beings is almost without exception harmful. Nevertheless, the nuclear industry again and again tries to play it down in order to have an "alibi" for increasing the radiation level caused by nuclear power stations and the discharge of the most dangerous fission and corrosion products into our environment. In the course of millions of years organisms have "adapted" themselves to natural background radioactivity, so that the percentage of individuals injured genetically or in health by it has been eliminated by the process of natural selection; as a result of this the species has remained healthy and has even been able to develop. But the positive forces of natural selection are largely ineffective today because human

70

civilization has given medicine the opportunity of keeping sick people and babies with genetic damage alive. This means that genetic defects will be inherited. According to Professor Bechert (11), it is quite correct to say that radiation from outer space affects us all. But the important question is how much extra radiation man can stand in addition to this unavoidable radiation before a "genetic catastrophe" occurs— when so many people are genetically sick that the healthy ones who remain have to make caring for the sick their major occupation. Research workers on genetics are not sure how much more of this damaging radiation we would be able to stand.

Attention is often drawn to the fact that in Kerala, in India, there is a whole population of 10,000 people living in a very high level of natural background radiation of 1300 mrem/year (129) (as against 130 mrem/year in Switzerland, for example). Professor Hedy Fritz-Niggli of Zurich argues that "these people, living under the shadow of high natural background radiation should be examined for radiation damage because no definite traces of this have been established to date." (34). This sort of statement tends to create the impression that the whole range of radiation levels from 130 to 1300 mrem is quite harmless.

However, firm conclusions should not be drawn from these proposed tests. For example, UNSCEAR states: "Careful, protracted study should be made of those groups of individuals which are, or have been exposed to high levels of radioactivity, such as the irradiated populations of Hiroshima or Nagasaki, people living in areas where natural irradiation is high and individuals who have been irradiated for medical reasons. Appropriate methods should be devised to extract from these studies all the relevant information on hereditary radiation damage that they can provide.

71

"An understanding of the hereditary effects of ionizing radiation cannot be obtained without a thorough knowledge of the factors affecting the establishment of hereditary traits in the population and above all, the pressures of mutation and selection and the genetic structure of the population. Continuous, large-scale investigations of human population groups living in different environmental, social and cultural conditions will have to be undertaken if the respective roles of the various factors involved are to be isolated."

This shows that the knowledge we have now is still totally inadequate to predict with responsible awareness the harmful effects of increased levels of radioactivity on future generations.

Furthermore the ICRP clearly assumes that populations living with high natural background radiation levels are likely to suffer greater damage. It reports (58): "The limit for natural background radiation for the great bulk of the world's population is of the sixth order. In a few areas of high natural background radiation the risk is of the fifth order."

This can only mean that the risk in Kerala, for example, from high natural background radiation is ten times greater than it is for the rest of the world's population. The argument quoted above—that no definite conclusions have been established about Kerala—therefore offers no guarantee of safety whatsoever. According to the ICRP natural background radiation is just as harmful. Statistical genetic investigations and the comparison of quite differently constituted populations are certainly very difficult. But life goes on in spite of statistics. If people are being harmed, although it cannot be demonstrated statistically, they are nevertheless being sacrificed. Because genetic consequences generally will not be seen or felt until they emerge in future generations, it is impossible to recognize them now. Obviously

72

the future does not lend itself to categorization and statistical evaluation. It is simply not known what symptoms will appear and to what extent. The harmful effects of radiation which will be only too obvious to society in the future cannot be predicted today.

Since it has been confirmed that natural background radiation is already having harmful effects there is no argument in favour of some sort of additional threshold radiation dose—rather the reverse! UNSCEAR writes: "At the same time, the exposure of mankind to radiation from an increasing number of artificial sources—including the world-wide contamination of the environment by long- and short-lived radionuclides from weapons tests—warrants the utmost vigilance particularly because the effects of any increase in radiation exposure may not appear for several decades, in the case of genetic disease, and for many generations, in the case of genetic damage.

"The committee therefore emphasizes the need for all forms of unnecessary radiation exposure to be minimized or eliminated completely, particularly when the exposure of large populations is involved."

That the responsible authorities should fail to recognize the force of such unequivocal statements is really enough to make one despair. It is more than evident that the existence of natural background radiation cannot be advanced as an argument to justify radioactive emissions!

At this point an indication should be given of the amounts of background radiation that are involved. Cosmic radiation, which produces secondary radiation through atmospheric changes, produces a level of about 50 mrem/year at sea level. However, this amount is subject to fluctuation depending on its neutron components and on an uncertain QF factor.

Radiation from the earth is usually of about the same order. Radiation is produced primarily by the minerals uranium 238, thorium 232 and radium 226. In a few areas, particularly where radioactive gold is to be found, the level of radioactivity is far higher. There are places like this in Brazil, Niue Island, and in India in Madras and Kerala. However, only the latter two places have high population densities (129).

It must be made clear that in the above-mentioned regions the external radiation is abnormally high and is primarily gamma radiation. However, this does not mean that such amounts of radiation can be compared automatically with the emission of highly dangerous artificial radionuclides from nuclear power stations, because these latter substances can accumulate in the body and in the environment.

Attempts to prove by comparison with background radiation that nuclear power stations are harmless can therefore be misleading. Nevertheless this trick is often used. The following comparison provides an example of the lengths to which apologists for nuclear power stations are prepared to go (3): "Anyone moving house to a place 100 metres above sea level is increasing his background radiation level more than he would by moving into the vicinity of a nuclear power station." In this statement no account whatsoever is taken of the fact that by moving house further above sea level only the gamma radiation is increased, while an atomic power station emits a highly dangerous mixture of artificial radionuclides which penetrate the waterways and accumulate in the soil, the air and the biomass. So the comparison is innapropriate and unfairly seeks to minimize the dangers to an unsuspecting public.

Furthermore the calculation of risk is based on an additional radiation burden from the reactor of only 1 mrem/year! The radiological protection laws allow for a radiation level of 170 mrem/year for the world population. The calculation of risk must obviously be based on

74

permitted limits (although these ought to be lowered consistently) and not on some sort of average which can only be verified with the greatest difficulty. The ordinary citizen has to rely on what the law allows.

The Swiss Atomic Energy Commission, in a newspaper announcement (112) in May 1971, triumphantly declared that the holders of all the professorial chairs in medical radiology in Swiss Universities had come to the unanimous conclusion that "the risk of an additional 1 mrem of radiation in the vicinity of a nuclear power station carries no significance for the populations living there and for other forms of life, in comparison with the variations in amounts of natural background radiation, or with the risks from the medical applications of ionizing radiation." This is in the end a confirmation that natural background radiation is significant and harmful!

The newspaper article quoted above does not do justice to the real problems caused by radiation from nuclear power stations and the peaceful application of nuclear energy. For example, not enough attention is paid to potential biochemical damage by mutation (see p. 110). The extra radioactivity released into our environment by reprocessing plants is a problem that inevitably belongs to nuclear power stations themselves, for without these plants the power stations would not be economically viable. Furthermore, the value of 1 mrem is not relevant to the argument, which should be concerned rather with the additional radiation dose allowed by law, i.e. 170 mrem/year for the peaceful application of nuclear energy. The holders of the professorial chairs very prudently make no mention of the crucial 170 mrem. Before long this 170 mrem (and indeed the 500 mrem in the vicinity of a nuclear power station) will no doubt be made to appear a quite trivial amount and will be labelled as harmless.

However, as recently as May 1971 Professor Sternglass

announced that he thought that he could prove that an additional dose of fallout (from nuclear weapons or nuclear power stations) of 1 mrem can increase infant mortality rates by $\frac{1}{4}$ per cent. He also came to the conclusion that statements first made by I. M. Moriyama might well be correct. According to Moriyama, not just infant mortality, but chronic genetic diseases as well, were much more strongly influenced by lower levels of environmental radiation than could be assumed from the results of experiments with high levels of radiation, performed on laboratory animals.

In Switzerland the permitted level of 500 mrem in the vicinity of nuclear power stations was not lowered to 170 mrem until the spring of 1971. This permitted level will probably have to be reduced even further as a result of pressure from groups sponsoring radiological protection research and the activities of opponents of nuclear energy —although the atomic industry so far has absolutely refused to investigate their demands. (By way of comparison, Gofman and Tamplin demanded that the permitted level be lowered to 17 mrem, see page 121.) If one considers the extent and variety of proposals, from many different sources, about the practical effects of radiation levels in areas where there are nuclear power stations (see page 84), one begins to understand the earlier insistence on high limits. With drastic reductions in the levels of radiation permitted, major alterations would certainly have to be made to a number of installations.

Furthermore, nuclear power stations cannot be considered in isolation (as we noted on page 75). The peaceful application of nuclear energy holds dangers for mankind in the shape of the *reprocessing plants* for irradiated fuel rods, to say nothing of *the transportation of atomic waste and the unsolved problem of ultimate storage*. The whole process of the peaceful application of nuclear energy must be considered, right from the excavation of uranium in the uranium mines. Highly radioactive rocks are found on the surface of uranium mines; they can disintegrate and the resulting radio-nuclides can then reach the food chain, although they might have remained undisturbed without the mine.

In conclusion, we should look at the combined effects of natural background radiation. In all, a level of 126 mrem/year is produced by natural background radiation. This level is related to total body irradiation.

External radiation:
Cosmic radiation	50 mrem/year
Radiation from the earth	50 mrem/year
Internal radiation	26 mrem/year
Total natural burden	126 mrem/year

4 · ARTIFICIAL RADIOACTIVITY

Man first felt the increase of artificial radionuclides in the biosphere at the end of the Second World War when the atomic bombs were dropped on Hiroshima and Nagasaki. The nuclear products from these explosions were disseminated throughout the whole world. From the atmosphere, they reached the soil and the water, were assimilated by all forms of life and ultimately reached mankind in food. The distribution of these individual radionuclides in the biosystem is extremely diverse, depending on their physical and chemical characteristics. Amounts of nuclear products, or their half life, are not the only decisive factors to take into account in assessing their danger. Equally important factors are: the selection processes in plants and animals; the resorption capacity of the accompanying chemical elements through the organisms; where they accumulate and the time it takes them to accumulate in particular organs of the body.

Artificial strontium 90, which can become concentrated in the bones, is particularly dangerous. Its 29 year half-life is relatively long. It does not disperse evenly in the bone structure—because of multiple increases in radiation it gives rise to "hot spots" depending on the supply of calcium. Nobody can predict where these spots will occur. Cesium 137 is also one of the most dangerous cumulative isotopes. This becomes concentrated in muscle tissue.

Because of atomic bomb tests towards the end of the fifties there was a world-wide increase in the radiation levels in plants and animals. The strontium content in infants' bones rose considerably, because milk is particularly susceptible to heavy strontium contamination as it contains so much calcium. This has doubled the effects of natural background radiation in the growing bones of young children. Eskimos showed 10–40 times greater cesium concentrations than normal because the reindeer that provide them with food had accumulated too much cesium in their muscles. The accumulation was the result of the reindeer eating lichen which is particularly absorbent.

After the reduction in nuclear bomb tests, the contamination fell off but then rose again as a result of the temporary resumption of Russian tests. The strontium 90 level in bones can be used as an almost direct indicator of such tests. Strontium is ingested by grazing animals and then passes into milk. Milk and cereal constitute a major source of food, and corn, like milk, is especially threatened because fission products tend to accumulate on its outer layers. In 1960 and 1961, in Germany, important foodstuffs became noticeably contaminated with cesium 137 and strontium 90 as a result of atomic bomb tests. Sections of the population were being fed 70 per cent of the permitted permanent level in their daily food. As a result the German Ministry of Agriculture had to come to grips with the questions of whether the sale of wholemeal flour and black bread should be prohibited and whether the full scale milling of corn would have to be limited. As radionuclides accumulated particularly heavily on the outside layers of the corn, which are richest in vitamins, the situation might almost have been reached where the most nourishing parts would have been compulsorily removed because of contamination and replaced with less healthy white flour.

But the full implications even of these serious dangers from radioactivity were never fully made clear, but minimized. Professor Hedy-Fritz-Niggli writes about fallout (34): "If one looks at the additional artificial radiation resulting from world-wide fallout from nuclear devices which have already been exploded, then one comes to the surprising conclusion that (until 1962) such world-wide contamination only contributed 11 per cent to the natural background level, which is fixed at 100 per cent." However, such a comparison of the total fall-out from atomic bomb tests with natural background radiation avoids the real issue. This argument, based on the premise of general radiation levels, will obviously lead the layman to accept the percentages and to conclude that things were not really so bad after all. Fortunately, those involved in radiological protection research correctly recognized the acute danger from the concentration of radionuclides in the biomass, so that even the great and the mighty of this world were eventually forced to listen, and atomic tests in the atmosphere have been largely discontinued.

The dangers of nuclear power stations are being minimized in a similar fashion today. Using percentages or mrem, comparisons are made between the discharge of extremely dangerous radionuclides, which become concentrated in the environment and in the biomass, and the level of external background radiation natural for man (see the example about moving house on p. 74). This tends to deceive people and it is therefore extremely important that the public should be offered some sort of factual explanation. A population that is correctly informed would never accept the risks expected of them at the moment. Moreover, there are risks involved which cannot even be estimated.

Atomic bomb tests and the resultant expansion of nuclear technology caught radiobiology in a state of underdevelopment. It has not yet made up this leeway because it faces enormous problems, the main causes of which are: the length of the human generation period; the long latent period before the outbreak of radiation diseases; and the difficulties of carrying out mass experiments in the field of small radiation doses. However, as a consequence of this a new kind of science has developed, so-called *radio ecology*. This is a branch of radiobiology and is related to ecology and biochemistry; it is concerned with the relationship between a radioactive environment and the organisms that inhabit it. It is a colossal task to follow up the distribution pattern of the different radionuclides in the biocycle, particularly as the path, place and time of accumulation can depend on a large number of complex factors.

Similarly, knowledge of the biological effects of radioactivity has also been extremely limited. Present experience of risk is mostly based on atomic bomb explosions, which have served as a real-life mass experiments. The path of fission products could be traced from these explosions: from the rain into the waterways, into the ground, plants and animals. Soil, depending on its constitution, proved itself a good filter for radionuclides, so that underground waterways in fact contain very low levels of them. Conditions in the surface waters were also explored. Dr Manfred Ruf (97) has published a very comprehensive work in this field and has confirmed that the bio-accumulation of radioactivity has reached a level where fish, mussels and crabs are no longer edible, although the radionuclide concentration of the water itself still permits its use as drinking water.

The following concentration factors[1] for cesium 137 were discovered:

seston	10–13,000
green algae & moss	200–7,300
pond weeds	200–1,800
sediments	100–17,000
fish, muscular systems & gonads	40–9,600

Detailed statements exist on the concentration factors of the different fission products (97).

It must be stressed that the individual concentrations of radionuclides produced by radioactive fall-out are still well below the maximum permissible concentrations laid down by the ICRP for radiation levels for drinking water.

Although a great deal of evidence has already come to light concerning the distribution pattern of radioactive deposits in the biosphere, there are still gaps. Added to this, in the peaceful application of atomic energy unknown factors can always arise.

Dr Walter Herbst of the Institute of Radiobiology at the University of Freiburg in Breisgau writes (44): "The evidence from atomic bomb tests is certainly significant, but it does not relate to the additional risks of the use of

[1] The concentration factor is an ecological scale for measuring the extent of concentration processes. By it is understood the correlation between the activity of a radionuclide in the bio-mass and the concomitant water activity:

$$\text{concentration factor} = \frac{\text{Activity in water}}{\text{Activity in bio-mass}}$$

(depending on the unit of weight.)

atomic energy for peaceful purposes, for such risks are varied and diverse in many crucial respects. The classification of the level of risk, both of individual fission products and their constituent parts in compounds, varies. Other neutron-induced,[1] radioactive isotopes can have a critical biological effect. The radio-ecological processes diverge considerably. The type of radiation load, in the long term, is specific to each case."

Nuclear power stations therefore pose further problems. When they are operating normally they load our environment with radioactive discharges. In particular, noble gases escape into the air and of these Krypton 85 and xenon isotopes are especially dangerous because of their long life span. In the past there was no Krypton 85 in the atmosphere. It is heavier than air and becomes concentrated close to the ground. It also penetrates the body through the lungs and bronchia and can occasionally accumulate in the fatty tissues and body fluids. According to Dr Herbst (44) it can be calculated "that, if present tendencies and techniques are maintained the whole world and the whole of mankind will become so heavily contaminated with fairly long-lived noble gases, especially Krypton 85 (physical half-life 10 years), in the coming decades that, through this alone, the present maximum tolerance dose will be largely exhausted."

Naturally the output of radioactive substances from an individual power station should conform to the radiological protection laws which, until now have allowed an additional load of 500 mrem/year in the vicinity of a nuclear power station. It is quite incomprehensible that legislation should permit such high doses and that reductions have always been stubbornly resisted. The

[1] Through bombardment with neutrons, which are of course produced in nuclear fission, non-active material can also become radioactive. We then talk of induced radioactivity or activation— or corrosion-products.

Swiss atomic energy industry officially states (33): "In the vicinity of nuclear industry additional loads were established that have so far never exceeded 1 per cent of the permissible radiation level of 500 mrem/year, on average 1 mrad, therefore about 1/100th of the amount of natural background radiation."

In a recent publication (80) (1970), on the other hand, two well-known German atomic experts and supporters of nuclear power stations write: "The radiation effects on people employed in nuclear power stations are less than those laid down for professional radiologists. Nobody anywhere in the vicinity of a nuclear power station will be even temporarily subjected to more than a quarter of the natural background radiation (30 mrem/year) as a result of the discharge of radioactively contaminated waste gases or radioactive effluent."

For the construction of the largest European atomic power station, Biblis, a few miles north-east of Worms, it was announced that the radioactive level in the vicinity of the power station would amount, at the very most, to 20 mrem/year (30). Westinghouse Electric's advisory environmental expert is supposed to have summed up operational experience with nuclear reactors by explaining that, for the population in the vicinity of a nuclear plant, the radiation level would only reach 0·002 mrem/year, compared with 120 mrem/year from natural environment radiation (79). Furthermore, the USAEC (American Atomic Energy Commission) maintains that the average radiation level for the whole American population with all the existing reactors only amounts to 0·001 mrem/year (105). However, concrete figures are available on the operations of particular reactors. So Professor Sternglass (105) reports that the normal functioning of the commercial nuclear power station "Dresden" (a boiling water reactor) in Chicago, in 1967/68 produced a level of 114–350 mrem/year under the exhaust plume, at a distance of 3/5ths of a mile. The radioactive plume could be identified at a distance of 9 miles, at which point it produced a radioactive level of 22 mrem/year.

This hotch-potch of various data, average limits and temporary levels should be set out once again here:

0.001 mrem/year	Average level of radiation from all reactors for the US population, according to the USAEC. (105)
0.002 mrem/year	Maximum dose in the vicinity of a power station, according to the experience of Westinghouse Electric. (79)
1 mrem/year	Maximum yearly average dose in vicinity of a nuclear plant, according to the Swiss Atomic Energy Commission. (33)
5 mrem/year	Highest ever measured dose in the vicinity of a nuclear plant, according to the Swiss Atomic Energy Commission. (33)
70 mrem/year	Maximum level in the vicinity of a nuclear power station, according to German experts. (80)
20 mrem/year	Maximum level in the vicinty of a power station, Biblis (expected). (30)
114–350 mrem/year	Measured (i.e. calculated) in 1967/68 at the "Dresden" reactor, at a distance of about one mile, directly under the exhaust plume (according to Professor Sternglass (105) from a publication by the Bureau of Radiological Health of the US Department of HEW).

Well, which of these statements is true? The credibility of such crucial data is rather shaken by the diversity of these opinions. The population cannot prove anything because discharges of radiation cannot be seen or felt.

As larger and larger reactors are being built for commercial and economic reasons, and as their numbers are increasing rapidly, we can expect a rise in environmental contamination, as long as technological processes remain the same. All calculation of risk therefore must begin with permitted doses, particularly as these doses are higher than the ones that are possible with present technology. The starting points should not be measured or calculated data, because these vary according to their origins.

Very surprisingly, the special dose of 500 mrem/year for people living in the vicinity of a nuclear power station was removed in Switzerland, in 1971, and replaced with the one for the world population (94) which amounts to 170 mrem/year. However, even here it is quite incomprehensible to the uninitiated that limits should be set at 170 per cent higher than is necessary. In the last analysis it must be recalled that damage can begin with a radiation dose of zero.

The physicists, engineers and builders of nuclear plants boast that they have kept below the permitted dose (which they incorrectly take to be harmless or, at the least, only "required") by 170 per cent and are proud of their achievement. In fact the opposite should be the case; tolerance doses should be reduced at least to a level at which they can be vindicated. Power station designers might then feel encouraged to undertake even further improvements. The tactic of permissible high doses, which has been stubbornly justified for the last twenty years, leads one to suppose that there is no alternative. Under pressure from radiological protection research and

85

from opponents of atomic industry, it will certainly not be justified for very much longer.

Light is shed on the "secret" of the incredibly high doses permitted, if one knows that radioactive discharge depends decisively on the leakage rate of the fuel rods (see p. 66). However, this leakage rate cannot automatically be controlled, and it should be remembered that a reactor can contain up to 30,000 such rods—the bigger the reactor, the more rods.

Because the final warning systems built into the waste gas pipes are adjusted to the so-called "permissible" limits (121), it is obvious that the legal requirements can be fully exploited, The most recent technique of retaining waste gas deposits in storage devices of even greater capacity does nothing whatsoever to alter this. Working with such high "permissible doses," (that is to say high threshold doses) in accordance with the principle of minimalization of effect is quite unforgivable and serves only to deceive an ignorant public.

In the light of the possible development of our civilization and technology, even a sound minimalization principle, conforming to technological and economic possibilities, is strictly to be avoided for such dangerous emissions as radioactivity. *Primary attention must be paid to the protection of life and the environment, and any emissions of radioactive substances whatsoever from industrial power stations must therefore not be tolerated. In the present situation however, nothing at all can be achieved or altered by tight controls and extreme safety measures.*

Nevertheless, nuclear power stations do not disseminate dangerous substances into our environment solely in the form of waste gases. The effluent produced is just as heavily contaminated with active waste, even after purification. This includes activation and corrosion products, as well as tritium (radioactive hydrogen) in addition to the fission products already mentioned. However small the concentrations in the effluent, they can still accumulate in the bio-mass and in sediments. This fact, however, still does not appear to be generally recognized. Dr H. R. Lutz, works manager at the Muhleberg atomic power station in Berne, writes in one of the big daily newspapers (81): "The operator of a nuclear power station must guarantee that (as with waste gases) the concentration of radioactive substances in the effluent leaving the plant is so small that any radiobiological threat to the public is entirely eliminated. For example, they should make certain that effluent could be used as drinking water without risk. Who could say that of the effluent leaving industrial plants? Fears that in spite of the low activity output certain radioactive substances could become concentrated in the bio-cycle have been proved groundless."

On the other hand, a publication by the government adviser for chemistry, Dr M. Ruf of the Bavarian Biological Research Unit in Munich (96) may be quoted. According to Ruf, the flow of radioactive effluent into open water should not be seen as a physical dilution problem, in which dangers are only to be expected if the recommended limits laid down for drinking water by the International Commission for Radiological Protection are exceeded. Instead the dangers lie rather in the possibility of radionuclides becoming concentrated in the bio-mass.

For example, the following concentrations of fission and corrosion products which arise in nuclear power stations have been found in fresh water (96):

	Sediments	Phyto-Plankton	Water Plants	Fish
Strontium 90	10–500	10–1,000	10–10,000	1–200
Cesium 137				
Cobalt 60				
Iron 59 etc				
		200,000	100,000	

D

Radiological research has recognized these dangers and, if the behaviour patterns of radionuclides in water are known, some attempt to protect the public can be made by establishing standardized limits of radioactivity for rivers. With the increasing number of atomic reactors, all these problems are becoming more and more significant.

Furthermore, there is also the danger of the contamination of fruits which are commercially watered or fertilized with sludge. A further problem, which is often ignored, is the eventual hereditary damage to members of our eco-system and the effects resulting from such damage. It can be expected that even a very slight rise in the radioactivity level in the waterways will lead to a higher mutation rate in fish (97).

Finally we should quote G. Harringer, engineer and ministerial adviser to the Ministry of Agriculture and Fisheries in North Dusseldorf, who, commenting upon the pollution of the Rhine water through nuclear energy plants, stated: "Nuclear energy will be applied increasingly in trade and industry, medicine and research. It is quite beyond doubt that it will provide a large part of the future power supplies and will be used for industrial process heating. In the Rhine drainage area, twenty nuclear power plants are at present planned, not to mention the acclerator plants envisaged. On top of this, there exists the fallout from nuclear weapons. It is therefore possible that not only the moving waters, but also the plants, animals, and mud in the waters will be contaminated. Such contamination could begin to cross national frontiers thus leading to an international problem."

Such statements by conscientious scientists indicate what sorts of dangerous consequences the potential accumulation of residual activity in reactor effluent can have. It is absolutely essential that the public should be fully informed of these unpleasant and as yet totally unexplored facts.

5 · THE BIOLOGICAL EFFECTS OF RADIOACTIVITY ON HUMAN BEINGS

Anybody who wishes to discuss the risks of nuclear power stations or form a well-founded, objective opinion, must first survey the possible biological effects upon our health. This can best be achieved by reviewing aspects of relevant radiation dangers in the context of current radiological protection research. The majority of the following arguments are based upon the publications of the ICRP, that is to say, the task groups it has set up, and the UNSCEAR. Both these committees are entirely independent of the nuclear power industry and cannot be influenced by it. Particularly important is the fact that radiological protection laws are based largely upon the recommendations of the ICRP.

In any consideration of biological radiation effects, a clear distinction must be made between damage to health, (also called *somatic damage*), which affects the irradiated person himself, and *genetic damage*, which does not become apparent until later generations. Somatic damage is further sub-divided into *acute radiation damage*, which can occur in connection with nuclear accidents or catastrophes resulting from high radiation doses, and *delayed damage*. The latter can also arise as a result of the short or long term effects of small doses, whether internal or external.

It must be stated quite unequivocally that radiation damage begins from a zero dose upwards. This was expressed by the UNSCEAR as follows (131): "The study of the relationship between dose and effect at cellular and sub-cellular levels does not give any indication of tolerance doses and must lead to the conclusion that

89

certain biological effects follow irradiation, however small the dose may be." This report goes on to explain that there are still several complex factors determining whether primary radiation influence will appear in the form of immediate or delayed damage. The particular sensitivity of the individual human being is one of these factors, and this is not considered by radiological protection laws, which is the same as saying that the sensitive individual must be sacrificed!

(a) Acute damage

Acute radiation damage as a result of total body irradiation with medium and high doses could only be studied after the atomic bomb explosions in Hiroshima and Nagasaki in 1945. Apart from burns and other apparent injuries, the victims also showed symptoms of the acute *radiation syndrome*, which varied according to their individual sensitivity to radiation. The individual symptoms, which become evident in this condition after hours or days, can also be observed in certain other diseases. The people in this instance suffered from headaches, dizziness, vomiting, fever, and diarrhoea, became apathetic, and often died after a few days. But even those who were apparently uninjured suddenly became ill after one or two weeks, vomiting blood, displaying bloody spots all over the body and haematuria; later their hair fell out and blockages and feverish infections occurred, which resulted from a lack of white blood corpuscles. Survivors faced life-long infirmity. Later accidents in the exploration and application of nuclear technology then provided opportunities to confirm and expand a picture of the acute radiation syndrome.

The effects of a single total body irradiation are something like the following: With 0·25 rem, no effects are evident; with 25–60 rem, 10 per cent react with vomiting and nausea; with 180 rem, the first deaths occur, and 25 per cent of those irradiated suffer from radiation diseases; with 300 rem, there is a 20 per cent mortality rate, and with 420–700, a rate as high as 90 per cent. With doses over 1000 rem, survival is unlikely; death normally occurs after a few hours. However, it should be noted that the data in the literature shows no uniformity in this respect.

(b) Delayed damage

It is not only these large radiation doses that are dangerous, however, for smaller and even the smallest doses can lead to so-called delayed damage. This is not a question of specific illnesses but rather of effects which can occur as a result of damage caused by other known and unknown factors. They only reveal themselves after a certain dormant period, possibly after years or decades of irradiation or permanent exposure to small doses of radiation. If germ cells are also involved, one then speaks of *genetic damage*, which only appears in the affected person's children or in later descendants.

(c) Aspects of radiological protection

What sort of foundations are there to support contemporary radiological protection practice? What kind of scientific evidence is at our disposal?

If one were to believe the atomic energy industry, then everything in the garden would be lovely. Thus, for example, we read the following in the *Neuer Zurcher Zeitung* (32): "The International Commission for Radiological Protection has established norms for maximum permissible irradiation which, if properly heeded, will protect against every kind of genetic effect or threat to health."

The majority of the population are certainly labouring under this delusion, even though it is entirely false. The establishment of maximum permissible irradiation norms

91

depends upon how many people should be sacrificed from a particular population group—for people will certainly die later as a result of high permissible radiation levels. They will die from such delayed damage as will result in cancer or leukaemia. The ICRP expresses this in these terms: (66) "Where tumours and genetic effects are concerned, it is generally postulated that no threshold exists. Recommendations insuring protection now have to be designed to reduce the total probability of such effects to an acceptable limit within a population."

But this is not as simple as it sounds, for the corresponding risks can never be calculated quantitatively simply because the scientific data are lacking. Apart from that, (because the crucial elements of radiobiology still lie largely in the dark), it is also impossible to estimate today the genetic effects on future generations. Mutation percentages alone tell us far too little. In the following, we will try as far as possible to shed some light upon this darkness by reviewing the loopholes in our scientific knowledge.

In this latter respect, the ICRP is absolutely frank. It describes the problem in a recent publication as follows (55): "The foregoing makes it apparent that this report can provide no simple solution to the practical dilemma of setting criteria for radiation protection. Although quantitative recommendations are required for the control of the nuclear energy industry and for the protection of the population in emergencies, the evidence on which such recommendations can be based is imprecise."

To this honest statement by the ICRP can automatically be added, *that the bases are not only imprecise, but, in relation to crucial missing elements, entirely non-existent.* It is small wonder that the poor scientists find themselves in such a dilemma. We shall return later to discuss these crucial missing elements.

92

Radiological protection research is, fortunately, hastily preparing further necessary evidence. Recent research —also published by the ICRP—unfortunately leads us to believe that the risks have been misjudged again, which means that the present limits will have to be lowered even further. Mistakes have been made in this connection for years. It is not without reason that the definition of the "threshold dose" for human beings includes as many escape clauses as possible. The "threshold dose" is understood to mean that dose of ionizing radiation which, according to present evidence and experience, and including all safety precautions, will cause no added bodily harm to the person in question during his whole lifetime. It is significant that this dose (for those in employment) has had to be reduced over and again in small stages from 201 rad/week in 1904 to 0·1 rad/week today (82). Other so-called threshold doses—or "maximum permissible doses" (MPD), as they are called today —also had to be reduced and are also to be checked in the future. They are once again fluctuating—particularly sharply at the moment.

Even the current 0·1 rad is seen by many scientists to be at least ten times too high. The ICRP, in its 1969 publication, has already indicated in a detailed fashion the uncertainties of the present permissible dose and notes the following (68):

"The choice between no change and a partial or tentative change or revision (of the recommendations) will depend, so it seems to us, not only upon scientific proof, but also upon practical considerations such as the desirability of the general stability of the recommendations over a specific period of some years. The balance between practical considerations and incomplete scientific evidence is a matter for judgement outside the Task Group's terms of reference."

Because the ICRP has not recently lowered certain

radiation limits, it is not safe to say that everything is all right. Rather one is conscious that the whole structure of the ICRP's recommendations rests upon false premises (as will be explained later in more detail). The ICRP is already debating how to correct matters, but the new scheme for the calculation of radiological risk then becomes much more complicated. However, none of this new evidence can materialize today in any case, because essential elements are still unknown.

A further important consideration is that the current approved system for the implementation of radiological protection laws is, in practice, very simple to handle. The new, correct system would inevitably be much more complicated, which could give rise to difficulties in practical implementation.

Added to all this is the fact that the ICRP obviously does not want to chance recommending lower limits whose scientifically established bases are as imprecise or as non-existent as those at present in operation. Unfortunately, there also appears to be resistance to a reduction of the limits, even as a preventive measure for (as is quite clearly expressed in the above ICRP quotation) attempts are being made to maintain the whole structure of radiological protection regulations unaltered for as long as possible. Repeated alterations of such limits (which was the rule in the past) unfortunately give far too much publicity to the dangerous uncertainties which are inevitable in attempting to establish such crucial data.

However, mistakes today do not only affect the individual person or group but the population of the whole world, including generations to come. The task group of the ICRP describe this awkward situation not without reason as a "dilemma". The only correct solution would be to forbid absolutely the "peaceful" or rather *"pathogenic" application of nuclear energy*, because the

94

effects cannot be estimated precisely while scientific data are lacking. Quite contrary to this logical deduction attempts are being made to carry projects through by all possible means, so that the practice of radiological protection is forced into the service of technology, instead of serving the general good of the whole of mankind.

Dr Herbst describes this state of affairs as follows (44): "Mankind stands upon the threshold of revelatory progress in the fields of nuclear, isotope, and radio technology. The corresponding potential of actual risks grows accordingly Our deliberations and reflections occur at a time in which increasing offensives are being mounted by conscientious radiological protection researchers the world over, against the practitioners of conventional protection, who are being forced back more and more on the defensive."

(d) Correlations between the application and effects of radiation

It is appropriate to the rest of our discussion to trace some of the correlations between the application and effects of radiation. The biological effects are always dependent upon four factors:

the type of radiation, which influences the dose;
the dose response curve;
the temporal distribution of the dose (in some cases);
the spatial distribution of the dose upon human beings;

1 Type of radiation and size of dose

It should be quite clear, that the size of the dose has an important influence upon the biological effects. Indeed, we have already seen that, with acute radiation damage, the injuries become worse with increased doses. However, the inaccuracies inherent in radiobiology begin with the very definition of the physically absorbed dose in rads

95

D*

and the biologically active radiation dose in rem. These doses cannot be directly measured in the body. We can only try to calculate them with external physical measures. This is most simply achieved with X and gamma rays, where rad and rem are interchangeable. But as soon as particle radiation is added (with alpha-rays and neutrons), the Quality Factors (QF) must be included in the calculation of the biologically effective dose (rem = rad × QF). As a result of this the dose in the critical organ cannot be defined at all in many very important instances (such as internal radiation through unequally distributed radionuclides).

The establishment of dose size is also extremely problematic, and this involves the crucial load either for the individual or for the population in the vicinity of a nuclear power station. This dose is not only produced by external radiation but also by the artificial radionuclides which are discharged and absorbed by people through food and water, through the skin, and by inhalation, and can therefore irradiate the body from inside. Professor Tsivoglou comments in reference to this that (121) such body loads resulting from internal radiation are not, as a rule, directly measurable. To measure them could be undertaken only at great expense, with certain radio-isotopes in a whole body counter (a measuring chamber inside which the person in question sits). Failing this, one would have to estimate the body level *for each individual isotope* as it occurs in our daily intake (food, air and water), and this *would have to be based upon a whole series of estimates and assumptions.*

Naturally, the concentrations of the individual isotopes in the different environmental media (in every single kind of food, in air, in water, etc.) would have to be defined. In fact, it would be possible to carry out these analyses with great precision for individual cases, but anybody can see that the continual control of the additional level received by people in the vicinity of nuclear power stations would be quite impossible in practice. Moreover, nuclear power stations emit a diverse mixture of radio-isotopes.

However, in the operation of nuclear power stations, we are compelled to make do with procedures that are still very imprecise; this means that we are setting limits for waste gases and effluents before they reach the environment. Having reached the outside world, they will then be "minimalized" again—according to technological and economical possibilities—and the whole lot served up in suitable amounts to the population (which would obviously have to consist only of standard human beings).

The following schematic representation of the biological cycle of radionuclides shows how many estimates and assumptions have to be made, in order to deduce from the gross radioactivity given off into the environment by a nuclear reactor (the total discharge of radionuclides) the dose in mrem/year inflicted upon human beings:

RADIOACTIVE WASTE GASES OR EFFLUENTS

GROSS ACTIVITY (Ci/sec.)

Concentrations in the environmental media (food, water, air, etc.)

A series of estimates and assumptions lead to an evaluation of the absorbed dose for a standard human being.

Daily intake of radioactivity, which a standard human being will take in.

Body load through radiation

RADIOACTIVE WASTE GASES OR EFFLUENTS		GROSS ACTIVITY (Ci/sec.)
Radiation-absorbed dose (dose-size)	=	Dose (mrem/year) Because of totally inadequate radiobiological evidence, the health risks for specific dose sizes may only be roughly estimated.
		Health and genetic risks for the individual and for future generations.

Because of inaccurately ascertained dose sizes, the disease risks for the population can therefore only be roughly assessed, which means that they are not once examined in the light of exact dose sizes. Even the few spot checks now being carried out on the environmental media in the vicinity of a nuclear power station and at isolated measuring stations can change nothing in the present situation. Our lack of knowledge about small radiation doses will be discussed in detail below.

2 Dose response curve

Every disease caused by radiation is correlated with the amount of radiation involved (63). An example will clarify this. We can guess today that, of one million people subjected to a total body irradiation of the same dose (1 rem), 20 will become ill with leukaemia, as a form of delayed damage (52). Therefore, where small doses are concerned, not all of those irradiated become ill, only a certain percentage do. So we talk of a radiation-disease risk; generally the higher the dose, the greater the risk will be. Therefore for each particular dose size a quite clearly defined risk can be estimated for an individual, i.e.

a certain probability of becoming ill as a result of irradiation. In the example mentioned above, the risk would be 20 : 1,000,000.

This relationship between the disease and the radiation dose is represented by the *dose response curve* (62).

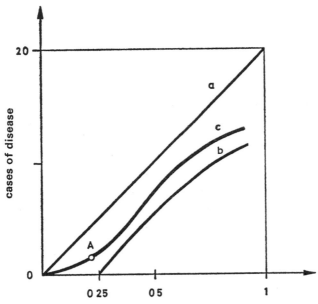

Fig. 3.

Figure 3 represents three possible dose response curves. The vertical line shows the number of cases of illness, ranging from 1 to 20; the horizontal shows the dose size increasing (arbitrary example) from 0 to 1 rem. From this we can read off the number of sick for each dose size. (The data must of course also include the number of people this represents, e.g. one million as in the above example of leukaemia.)

If a risk of disease existed of the order shown by curve *b* there would then be a real tolerance dose or, as it is called, a *threshold* dose, for this disease. Curve *b* cuts the

99

horizontal at a dose of 0·25 rem (i.e. not at the zero point with dose zero), which means that no cases of disease occur up to a dose of 0·25 rem. According to curve *b*, the tolerance dose for this disease is therefore 0·25 rem.

If we look at curve *a*, on the other hand, we see that this cuts straight through zero. Here there can be no tolerance dose, because each increase from zero causes more cases of disease. For example, at 1 rem we can read off the corresponding number of cases of illness on the vertical.

With curves like *a* and *c*, one can no longer talk of tolerance doses because each increase in the dose above zero causes cases of illness. Here we talk rather of *Maximum permissible doses* (MPD), whereby a certain number of cases of illness (that is to say, death resulting from delayed damage) are considered permissible as a consequence of a certain permitted dose of radiation! The ICRP today assumes that no tolerance dose exists for almost all radiation diseases, that is to say it assumes an *a* curve (66).

Curve *c* is a further example of another potential risk. We can see that at point A it comes down very close to the horizontal; therefore a tolerance dose almost exists, but not quite. From A it runs down very gently to zero which means that relatively few instances of disease arise in this area. Such a curve would be typical for a disease that affects a small but radiologically very sensitive section of the population (if it causes reduced fertility, for example).

In this dose response curve we see the scientific basis of radiobiology. An understanding of the scientific background is absolutely essential to a serious calculation of the risk of radiation dangers and also to the establishment of well-founded radiological protection legislation. However, we shall see, later, just how inadequate present-day knowledge is in this respect.

100

3 Temporal dose distribution

The temporal distribution of a radiation dose can also be important. We were accustomed to taking these conditions into account as a result of our experience with chemical poisons, and we hoped that the same conditions would apply to radioactive "poisons". So, for decades, we have clung to the concept of the threshold dose without having any proof of it. As a result of this the "tolerance doses" which were established were constantly proved wrong and constantly being lowered. There are true tolerance doses for many chemical poisons, because the body is often capable of producing quite harmless substances from them by chemical changes. The body can always control the introduction of small amounts of chemicals, so that damage does not necessarily result.

The whole of life depends upon chemical reactions. Ionizing rays, on the other hand, are fundamentally hostile to life and no organism needs them to live.

Radioactivity cannot be destroyed artificially or biologically, and radioactive substances continue to radiate for as long as it takes to transform all their atoms. An organism is capable of distinguishing them—according to their particular characteristics—but they irradiate the body nevertheless, even if their stay there is short-lived. After leaving the human body they can enter biological cycles again and cause further damage.

Atom experts who lack sufficient biological knowledge often dream up quite extraordinary possibilities. The comment: "Cooking salt is a poison too" has become famous in comparing the potential dangers of radioactivity in a nuclear power station with those of a large cooking salt factory. Even the production of cooking salt—so it was said—could kill many people, for only 3 grammes per 2 lbs of body weight would be enough to kill a human being. Everybody should see that such comparisons are irresponsible and devoid of logic, for the

101

body needs cooking salt in small doses for the maintenance of life, while it does not need radioactivity, which causes damage even in the smallest amounts. In contrast, cooking salt in small doses is essential to life and it has no cumulative effects. It is senseless, therefore, to try to play down the quite different and pernicious effects of ionizing radiation by comparing it with "poisonous" cooking salt.

The ICRP makes the general assumption that a single short term dose of radiation has the same biological effect as the same dose administered over a long period, either continuously or in small amounts. To take an example, this means that a radiation dose of 1 rem in one hour has the same biological effect as a dose of 1 rem spread over one year.

This argument is expounded quite unequivocally in the radiological protection laws for world population. At the moment they permit a dose of 170 mrem per year for the peaceful application of nuclear energy, in addition to the amounts of natural background and medical radiation. This amount was established primarily because of the gonad radiation limit which, according to the ICRP's thinking, should not be allowed to exceed 5 rem in the space of 30 years (30 × 170 mrem — 5 rem), i.e. up to the thirtieth year of life. We can see from this that an accumulation of the yearly dose is taken for granted from the very begining. But how was the 30 year period established?

Manstein (82) showed in the following example just how relentlessly and inhumanely maximum safety doses were established for the "standard human being". He says: "As far as the gonads are concerned, there is the particular danger of passing on even the slightest damage to the next generation. So statistics were called to the rescue. They stated that in the USA children were produced, on average, up to the thirtieth year of life. The threat implicit in this arrogantly schematic judgment is quite clear—woe to those who are out of step and who produce children later in life, for no account can be taken of them in present or future overdoses of artificial radioactivity."

Nuclear enthusiasts happily cling to the idea that, with certain kinds of genetic and somatic damage, the effect of long term irradiation in small or sporadic doses can be less than short term irradiation with the same total dose. How far the effects subside depends upon complex biological factors and on the recovery powers of the irradiated parts of the body.

Our present ignorance of the effects of irradiation in small doses should put us on our guard and we must not allow ourselves to be deceived by false hopes. There is too much at stake! For practical reasons, experiments could hardly be carried out in the field of small doses to try to establish statistically accurate results. Hundreds of thousands of experimental animals would be required and each one would have to be observed individually. (125). All the present evidence in the field of radiobiology for somatic damage with small doses, therefore, has been deduced from observation of the effects of large doses, which were administered in a very short time and only to certain population groups. (We shall be returning to this later.) Naturally animal experiments have also been carried out. According to ICRP practice, all the quantitative relationships observed with high doses are also valid for small radiation doses, which means that a linear dose response curve through the zero point is assumed (curve *a* on p. 99). UNSCEAR has this to say about these uncertainties (125): "Since so little is known about the effects of low doses of radioactivity great care must be exercised in inferring, from the available experimental or population-based data, the effects to be expected from irradiation from the artificial nuclides that are being released into the environment. While the importance of the very low dose of radioactivity to which they give rise may be great, it is difficult to evaluate, since their effect cannot be studied experimentally owing to the unmanageably large numbers of animals required."

These words from a quite unimpeachable scientific source illuminate the whole tragedy of what will inevitably occur if artificial radionuclides from nuclear power stations continue to be released into our environment. The consequences are not visible today because of the lack of scientific information on which to base judgement, and one does not have to be an expert or a scientist to be able to grasp this fact. An ability and a desire to read are the only necessary qualifications!

UNSCEAR writes further (126): "Short of obtaining adequate data on the frequency at low levels of radiation of such deleterious effects as leukaemia and other malignant diseases—and this will involve extensive human surveys and animal experiments—the use of any relationship to predict effects at low levels will, in fact, imply assumptions about the mechanisms which cause specific radiation injuries.

"In the present state of our knowledge any such assumptions would be largely speculative. The only justifications for comparing the effects of low levels of radiation with those observed for higher levels, and so assuming that there is no threshold at which malignant diseases start to appear, are expediency, and the consistency of the assumptions about the processes involved in both dose ranges. However, if we do compare low and high doses of radiation we do not know if in so doing the risk is underrated or overrated."

It is interesting to note that the UNO commission states here that, in the present state of knowledge, it is quite possible to underestimate the dangers of playing safe by establishing limits. As we shall see later even the

104

conditions observed with high doses are quite uncertain and based on shaky evidence.

4 Spatial dose distribution

In radiobiology the dependence of radiation effects on the spatial distribution of the dose in the body is a crucial factor. Generally it can be stated that irradiation of individual parts of the body or individual organs is likely to produce a lower risk than total body irradiation with the same dose. It can also be said that the effects of radiation become relatively smaller, the less the body is irradiated. However this conclusion is only valid if no account is taken of the differing radiation sensitivity of the various organs. It is this particular factor that up to now has been largely underestimated in radiobiology and radiological protection research.

In the early days of radiological protection there was just one maximum safety dose, also called the tolerance dose then. It was not until later that it was recognized that different body organs and tissues were not equally sensitive to radiation damage. Until the evidence for this was established, there was a great deal of suffering amongst those exposed to radiation at work. Even then a vociferous view maintained that there would be no danger as long as we keep to the norms established by science—and this is still the case today. Supervision was just as tight then as now—according to the contemporary state of knowledge.

In the following graphic example Manstein (82) demonstrates just how we can be deceived. In a watch factory, "The Luminous Watch Factory" in New Jersey, in the twenties, watches were produced with luminous dials. Hundreds of women were busily engaged in painting the numbers on the luminous dials with brushes, stroke by stroke. Luminous colours were used, which

105

were made from zinc sulphide with the addition of one ten thousandth of radium salt. The dilution was such that the amount of radiation that the women were exposed to lay well below the tolerance dose permitted at the time.

All this seemed to be fine but nevertheless more and more of the women workers were complaining of headaches and lethargy. Inflammation of the gums, maxillary suppuration and loss of teeth occurred and many of the women had to be hospitalized.

Research showed that the women had been constantly wetting the brushes with their lips, in order to keep them pointed and through this radium had been fed into their bodies continuously. It had carried on accumulating unnoticed and had irradiated the unfortunate women internally. Seven women died in unspeakable agony. The survivors did not even receive compensation, for the laws of those days did not recognize the concept of radiation damage!

It was wrong then to think that the danger had been overcome and it is still wrong, for the demons of artificial radiation are still stalking their prey today. However, today it is not "merely" a question of the threat to individuals who voluntarily expose themselves to radiation at work, but to whole population groups and now, indeed, to the whole of mankind.

Even today radiological protection laws are still based on the so-called theory of the critical organ. We have certain organs, bodily parts and tissues, which have been divided into four groups. For each group, a maximum safety dose was established (61):

	rem/year
Red bone marrow and gonads	5
Skin, thyroid gland, bones.	30
Hands, forearm, feet, knuckles.	75
Remaining organs.	15 each

These doses apply only to people exposed to radiation in their work. For other individuals, smaller population groups and for the world population other, lower limits have been established (particularly because of genetic damage) including reduction factors of 10 and 30.

In any case the acceptance of higher doses of radiation should only be expected voluntarily of people. Apart from the consequences of atomic bomb explosions this was always the case until nuclear power stations appeared.

In applying the above rulings it must be ensured that the named organs, tissues or bodily parts do not receive a higher dose of any kind of radiation than is prescribed. The organ on which the dose is based is therefore called the "critical organ". So, with total body irradiation the bone marrow and gonads are the critical organs, for they decide the permissible doses, because, with 5 rem/year they represent the lowest MPD. As a result of this people at work are not to be subjected to more than 5 rem/year (total body irradiation).

The whole structure of radiological protection, on which the recommendations of the ICRP are based, has recently proved itself both incorrect and unsuitable for the calculation of risk in the long term. It was based on typical wishful thinking and expedient optimism, and its fatal error was to disregard the risk of malignant illness to all those tissues which are not mentioned separately in the above list. This means that their relative radiation sensitivity is underrated by comparison with the bone marrow.

(e) The effects of radiation on cells and molecules
The influence on cell growth and on the constituent parts of cells (atoms, molecules) constitutes the most important category of radiation effects. Each organ is made up of billions of heterogeneous cells and the growth of biologi-

cal substances consists of the multiplication of these cell tissues through division. However, there are many tissues in the full-grown body which are constantly renewing themselves at varying speeds while old cells die off and new ones form (for example, white blood corpuscles, sperm, skin and peritoneum).

It was recognized very early on that radiation did not damage all the cells with equal severity but concentrated rather on those which multiplied very rapidly. The faster the rate of the cell growth in a tissue, the more sensitive it is to radiation. Sensitivity to radiation is therefore lower in an adult than in a child.

Sperm cells multiply particularly rapidly, which explains the high radiation sensitivity of the gonads (genitals) and that of the white blood corpuscles (leucocytes), which are constantly being formed and renewed by the bone marrow. Leukaemia caused by injury to the bone marrow has therefore become one of the most common radiation diseases. Fat, muscle, and connective tissue are the least sensitive.

Cancer cells grow very quickly, which is why medicine uses ionizing rays for their destruction. However, with such treatment a risk of radiation must always be taken into account, because healthy tissue is being exposed to exactly the same treatment. Naturally, in urgent cases, life-saving radiological treatment should not be avoided simply from an exaggerated fear of radiation damage. Nevertheless, the modern doctor has become much more circumspect in such cases.

Ionizing radiation also affects the chemical processes in cells. Molecules are split apart and radicals (fragments of molecules which can be chemically very active) and ions produced. These are very aggressive and form new compounds which are foreign to cells and can be poisonous. Even highly complex protein molecules can be destroyed

108

in the cell nucleus. Radicals can also be built directly into these proteins, which inevitably means damage.

Radiobiology is far too little aware of all these basic conditions, of which the UNSCEAR says (122): "All the essential constituents of cells and in particular complex molecules like proteins and nucleoproteins may be affected through the action of radicals. They may also be injured by radiation directly, however, without the intervention of radicals. The respective role of the direct and indirect action of radiation in bringing about cellular lesions is not yet clear; it is probable that in most instances, both modes of action operate."

This proves just how little modern radiobiology knows about the fundamental processsess of radiation-induced disease.

It becomes even more alarming when further potential damage to the cells is considered—called *transmutation*. By this is understood all those processes which are produced in addition to radiation effects as a result of the incorporation of radionuclides. This damage can cause atoms to change their mass and therefore give rise to modified molecular reactions. Moreover, with neutron emitters, stable atoms are radioactively induced, and finally with Beta-rays the risk is substantially increased, because the molecule in question becomes extraordinarily reactive and changes the elemental character of the atom.

In this way tritium 3 changes to helium 3, phosphorus 32 to sulphur 32 and carbon 14 to nitrogen. These radioactive atoms built into the molecule are therefore not only radioactive, but also able to destroy the molecule by their own transformations. Such an atom can also be built into a control molecule, by which process its biochemical and physical signification is changed.

The UNSCEAR writes of carbon 14 (123): "Since carbon is a basic constituent of all living structures, it has

109

also been suggested that the change of carbon 14 into nitrogen will sometimes occur in a key molecular structure; this change may add appreciably to the effects of the radiation released by that nuclide in the form of beta particles."

The unavoidable discharge of tritium 3 (heavy hydrogen) from nuclear power stations is highly hazardous. The nucleic acid molecule DNA, which carries genetic information, also contains so-called hydrogen bridges, which can be damaged by transmutations.

The macro-molecule of DNA is in fact shaped like a twisted ladder, in the manner of a double-helix. The components of the uprights in this ladder system are phosphate and desoxyribose, while various chemical alkaloids are built into the rungs in very definite ways. The opposite bases of the two uprights are linked together by hydrogen bridges. By replacing the normal, stable hydrogen atom 1 of such a bridge with the isotope, tritium 3 (heavy hydrogen), and through its transmutation into helium 3, which possesses little affinity, such hydrogen links or bridges are destroyed. Similar damage also results from the replacement of stable sulphur and carbon atoms by corresponding isotopes and their transmutations.

Dr Herbst describes the processes of tritium 3 transmutation and states (43) : "Because of the close connections which exist between the chromosomes and genetic damage on the one hand and the release of malignant neologisms on the other, tritium 3 must be expected to possess carcinogenous properties not only through radiation but also through transmutations. People believe they have been able to prove this by experiment— observations of health damage, for example, after the incorporation of tritium, which cannot be explained in terms of energy by simple radiation absorption, should also warn us to tread with caution. Especially worrying

110

is the fact that transmutation can give rise to much greater damage than the energizing radiation which accompanies the disintegration of the atom."

Unfortunately, the tendency still exists in radiological protection practice to pay far too little attention to delayed damage through incorporation. *The effect of transmutations is not accounted for at all in the present radiological protection laws!*

The common objection of experts who support atomic energy that, since time immemorial, there have been incorporated molecules in Nature, active carbon 14 for example, misses the essential point. As has already been explained, life has continued in a state of equilibrium with natural background radiation for millions of years because, in the context of natural selection, the detrimental effects were balanced out. However, this weeding out process is now largely ineffective for civilized man. There are also now many artificial radionuclides, which did not previously exist in Nature. Natural background radioactivity therefore *already represents an overdose* and in no circumstances should artificial radiation, be added to it. The consequences, particularly in the long term, are almost impossible to envisage.

(f) Hereditary damage

Genetic damage means damage to the genes and chromosomes. As far back as 1865, Mendel described two fundamentally different kinds of heredity: the dominant and the recessive. A mutation which is *dominantly* inherited is one that the child and all its descendants will transmit and the effects of the genetic damage it causes will be only too obvious. A *recessive mutation*, on the other hand, can continue to be inherited without obvious effects until it meets a similar mutation in the other parent; this can continue for many generations. Genetic

111

damage only becomes evident when the recessive mutation meets another of the same kind.

Our genotype is transmitted by complex molecules, the
so-called desoxyribonucleic acids (DNA), which have
already been described in connection with transmutations
and are to be found in the chromosomes. They are contained in the germ-cells and also in other body cells,
although here their numbers double. At fertilization the
mother's and father's chromosomes (23 each) make up a
new body cell with 46 chromosomes.

Mutations are the result of alterations in the structure
of the DNA molecules. They have always occurred but
the reasons for their occurrence are unknown. Energy-
rich rays can also produce mutations.

It is misleading for people to maintain that we understand radioactivity better than we do all the other poisons
which threaten us. Supporters of this view would like
us to believe that science has ionizing radiation "under
control". It is true that, unlike many poisons, it can be
easily detected because of its activity. But we do not
want to have to use geiger counters and complicated
machines all the time to test ourselves and our daily food
for radiation!

Now that the particular danger of radioactivity has
been recognized, it is all the more irresponsible to continue producing such proven mutagenous and dangerous
nuclear products in nuclear power stations, and to continue releasing them into our environment.

The relative ease with which apparently exact permissible doses can be established gives rise to a false sense of
security. The ICRP is quite clear on this question of
comparisons with other poisons (54): "The emphasis
given here to the limitations of the present assessment
should, however, not be regarded as implying that knowledge of the effects of low levels of radiation is any less
precise than that of the effects of other toxic agents or

112

environmental factors which cause similarly infrequent effects. In particular, it should be emphasized that the circumstances which introduce particular uncertainty into estimates of the genetic risks from radiation apply equally to the assessment of the long-term effects of all other mutagenous agents."

In comparison with other mutagenous poisons, detailed evidence about the effects of radiation damage is not nearly as advanced.

Of course, mutations must occur for higher life to develop at all. But the idea that genetic characteristics can be brought about through mutations is a false one. The effects are almost always negative and higher development can only be expected within the framework of natural selection. By this process live carriers of defective mutations are ruthlessly wiped out before they can pass on their defects. Mankind has every reason to continue irresponsibly raising the natural mutation rate because, as a result of the advance of civilization, natural selection can now do only part of its job. In spite of this an impression has somehow been created that the level of radiation permitted by the radiological protection laws for the world population, and the resultant increased mutation rate, are insignificant compared with natural processes.

Professor Dr Hedy Fritz-Niggli (34) writes in the *Basler Nachrichten* of 1 July 1970: "The natural mutation rate of genetic substance, the cause of which is not yet known, is already very high. What contribution is made by energizing radiation, the radiation of radioactive elements which is produced in the operation of nuclear power stations and could threaten us? As a result of current research it has been possible to estimate that, for example, 1 rad can produce two point-mutations in every 1000 sex cells, which can only mean that *1 rad (1000 mrad) produces one seventieth of the natural mutation rate*. With the inclusion

113

of all chromosome mutations it is estimated that *1000 mrad mutations produce in the region of 1/10th of the total natural mutation rate.* The maximum permissible amount of radiation (MPD) is established on a basis of possible genetic damage. . . . From the same point of view, the MPD for small population groups living in the vicinity of nuclear reactors was set at 500 mrem/year higher than the 170 mrem/year maximum for the total population. In fact, if these maximum permissible doses were to be reached each year, the natural gene mutation rate would be raised by 1/5th in 30 years."

As no further comment is made on these figures—apart from emphasizing that these limits are never reached in practice—the layman might be lulled into believing that there were still great reserves. However, mutation statistics often conceal a great deal. By themselves they state far too little about the detrimental effects which could arise in later generations and ultimately this is the crucial test.

UNSCEAR has the following to say about this (137): "For most genetic changes, we cannot even guess what the actual manifestations of damage might be through generations, in terms of individual or collective suffering."

The ICRP says, of the possible genetic damage to future generations: "It has to be hoped that the reader has by now no illusions about the difficulties of making predictions even of first generation damage. The arguments leading to the estimates are tenuous and the conclusions correspondingly uncertain. The problems of prediction of effects on future generations are even more difficult and the conclusions are even less certain. There is the additional problem of expressing the damage in meaningful units".(53)

"Because the total genetic damage will become apparent only over very many generations, it is only fitting

114

that the long-term effects should be a major preoccupation of the collective conscience; the 'total eternity damage' from this viewpoint may be considerable".(50)

From these quotations by the two top scientific committees in the field of radiological protection and radiobiology, it can be seen that present day science is totally helpless in the face of the long-term genetic consequences of the application of nuclear energy for peaceful purposes. This can only mean that the maximum permissible doses for today's world population can offer no security for future generations. The ICRP now thinks that the "collective conscience" should concern itself with this state of affairs. As no serious estimates of risk are possible today, it shows tyrannical arrogance to expect the world's population to accept a high level of radiation without adequate scientific bases and without frank explanations or requests for its opinion in the matter. The existence of nuclear power stations and reprocessing plants is indefensible. Even if they were set up in unpopulated areas, the radionuclides would eventually reach our environment. Krypton 85, for example, pervades the whole atmosphere!

(g) Somatic damage

In the field of somatic damage, too, we have to admit, with horror, that there is wide-ranging ignorance of the harmful effects of small permanent doses of radiation. Where does radiobiology's current, if meagre, evidence come from?

Almost all the evidence comes from surviving victims of the atomic bombs dropped on Hiroshima and Nagasaki during the Second World War. Since 1950 some 100,000 of these people have been observed and the causes of their deaths noted for evaluation. A certain radiation dose, calculated from the person's position at the moment of the

115

explosion, was attributed to each one. They were short-term, high intensity doses.

According to the doses attributed to the survivors, delayed damage has gradually developed and attempts are being made to provide quantitative correlations between dose and number of years elapsed before damage appeared for each disease (cancer and leukaemia). As it is a question of high doses here, which were administered in a short space of time (explosion), the limits discovered have to be interpolated according to the dose response curve (see p. 98). The correct relationship remains unknown, so a linear function through the zero point is assumed.

That all sounds splendid; but it is very imprecise. Why?

(a) The doses attributed to the victims are not exact. UNSCEAR writes (124): "Despite continued investigation, the doses received can only be inferred from the distance of the survivors from the hypo-centre of the bombs. The doses, therefore, are highly uncertain and this uncertainty reflects on the dose-effect relationship."

(b) The tissue sensitivity of the irradiated Japanese cannot be held as valid for the whole world population. The ICRP writes (69): "Even the Japanese population was not wholly 'normal': there was a lack of fit adult males absent on war service." UNSCEAR writes about this (136): "The surviving population has been heavily selected by the lethal effect of the irradiation itself, so that survivors may not necessarily be representative of the irradiated population with respect to sensitivity to radiation carcinogenesis."

(c) Furthermore, the Japanese were not equally irradiated for some of them were in houses, in front of walls, etc (133). One to two thirds of the survivors who received doses of 650 rad were more or less shielded (65).

Once again we can see a whole chain of uncertainties,

116

on which the radiological protection laws are based. A further important observation group is constituted by about 13,000 spondylitics, who were given therapeutic radiation treatment for arthritic diseases of the back in England in 1953/54. These people have also fallen ill with iatrogenous diseases over the years and such cases are being statistically evaluated in the same way as the Japanese. This evaluation is just as imprecise because the spondylitics were not "normal" people but were already ill; neither were they equally irradiated, with the result that many organs received unknown doses (65).

Finally, there are two more groups which have been studied. 1400 British and American radiologists were observed in the years 1897–1957 and 1938–1952. No increased leukaemia rate could be traced in the British; with the Americans, on the other hand, there were five times more cases of leukaemia than for the rest of the population. However, these observations betray great statistical uncertainties—particularly for the English— because the numbers of people studied were so small (70).

Practically all our evidence on delayed damage as a result of radioactivity stems from observation of the three above mentioned groups (surviving Japanese, spondylitics and radiologists). Besides these, however, there are other small groups, who have become sick as a result of delayed damage, because many diseases which are today treated by different means were previously treated with X-rays. Children suffering from swollen and diseased thymus glands were treated in this way. Doses of several hundred rad were used for this and such children would frequently become ill with cancer of the thyroid gland.

The insidious effects of radioactivity have also made themselves felt on the uranium mine workers. Even in the last century, when the effects of radiation were still unrecognized, it was observed that many miners fell ill with lung cancer during the excavation of ore from rocks containing large quantities of uranium; in the then badly ventilated, snowy mountain mines 75 per cent of the miners died. It is known today that the active inert gas radon, which is concentrated in mines, was responsible.

A quantitative statistical evaluation of the observations on the last mentioned small groups of people would be quite useless because the correlation between dose and effect (dose-response curve) cannot be provided.

To sum up, it should be noted that our radiological protection laws are based on observations of the three main population groups (Japanese, spondylitics and radiologists) mentioned above—that is to say, on very imprecise evidence. In addition to this, these groups had received high doses within short periods of time. Naturally animal experiments were also carried out, but mass experiments can hardly be performed for the crucial small doses, because hundreds of thousands of experimental animals would be required. So we have to use high doses and extrapolate to small ones, and this process itself gives rise to many uncertainties. So there is a whole chain of uncertain or unknown factors, on which our present knowledge of the effects of radioactivity in small doses is based.

6 · THE CALCULATION OF RISK

All the radiation-induced diseases that appear as delayed damage can also be caused spontaneously, i.e. without radiation. There are some 60 leukaemia cases per million people every year. There are even more cases of cancer.

Radiation risks are expressed in the same way. They are based on observation of the three groups of people already mentioned. According to the statements of the ICRP, it has been reckoned so far that the irradiation of one million people with 1 rad would cause 20 leukaemia and 20 cancer cases in the form of delayed damage, over a period

118

of ten to twenty years (55). The ICRP assumes that the risk is the same, whether the 1 rad dose is administered in a single irradiation or whether it is a question of continuing effects at correspondingly small radiation levels (55).

This rate of illness from cancer and leukaemia of 40 persons per million per rad may seem minimal, but it should be remembered that radioactive doses accumulate. If, for example, a person were exposed to an irradiation of 1 rad per year for 30 years then there would be a total radiation load of 30 rad (30 years times 1 rad). Because 40 cases of disease (cancer and leukaemia) arise per rad, a 30 rad radiation load (which has accumulated in the body in 30 years from a yearly dose of 1 rad) will produce $30 \times 40 = 1,200$ cases of illness.

In this way, the amount of illness arising from a particular dose size can be estimated—as long as the necessary correct evidence is available. As this is not the case, however, the ICRP has suggested expressing the radiation risk in rank numbers.[1]

The ICRP writes (51): "Estimates expressed in this way (e.g. 2 cases per year in a million exposed adults for each rad received) could, however, be readily misinterpreted as implying considerably greater accuracy than the facts justify. A fairer impression might be conveyed by defining 'orders of risk'. A fifth order risk therefore means that 10–100 injuries would be expected per million persons. Guidance in these terms would be sufficiently precise for practical purposes."

The last sentence is significant. We are more or less compelled to calculate in terms of hundreds and

[1] Risk   Cases per million people
6th order   1–10
5th order   10–100
4th order   100–1000
3rd order   1000–10,000.

thousands of victims and to say, for practical reasons, that this will have to suffice, simply because our calculations cannot be made any more precise. The ICRP is honest enough to admit this. Here, again, we can see quite clearly that it is quite indefensible, in the light of such uncertainties, to ask mankind to face the unquestioning acceptance of so-called radiological protection laws which have no scientific foundation. However this only concerns part of the danger to health. Genetic damage is very difficult to predict, as has already been explained.

Certainly radiological protection laws are necessary, in spite of their colossal loopholes. However, acceptance of the dangers of radioactivity should only be expected of those who take them upon themselves voluntarily from professional, medical or other reasons. Nuclear power stations, however, lead to an additional load for the whole population, particularly for those who live in close proximity to such plants. A correctly informed public would never accept as reasonable the risks that they are expected to face at the moment.

As has already been mentioned, the concept of the critical organ, on which radiological protection laws are based, has already been shown to be largely false and—particularly in the long term—unsuitable for calculating risk. According to this theory, the maximum permissible dose was established with reference to the bone marrow and gonads—until now thought to be the most sensitive organs (see table on p. 106). Disease of the bone marrow is certainly responsible for a common form of leukaemia which, until recently, was rated as the most common radiation illness. However the risk of cancer in many other tissues also susceptible to malignancies, has been largely ignored in these calculations.

Observations of the Japanese victims of atomic bombs confirmed these results. After leukaemia had taken hold initially in the first 13–15 years as expected, it then

120

declined from 1958 onwards and everything appeared to be fine. The number of remaining cancer cases was about the same as for leukaemia but then they began to rise quite unexpectedly. The ICRP has the following to say about this (70): "Cancer cases in excess of the expected number began to appear among the ankylosing spondylitics and in those Japanese who had been exposed to radiation, only 10–15 years after exposure, and it is impossible to guess when, or even if, the increasing number of cases will begin to decline."

According to the most recent research with the ankylosing spondylitics, it could even happen that the number of cancer cases could exceed those of leukaemia by a considerable amount. The ICRP writes (64): "The data in the table may also suggest that malignant diseases other than leukaemia will be 5–6 times more frequent than leukaemia when the results are assessed after 27 years of observation. However, in this context, the rates cited for 15–27 years after irradiation are quantitatively the most important and it should be stressed that these embody considerable statistical uncertainty."

The delayed appearance of cancer cases is explained by the longer latency period for the outbreak of cancer, compared with leukaemia. In 1970, a great stir was created by the writing of Gofman and Tamplin[1] who calculated, as a result of their many years of research, that the result of the present maximum permissible dose for the world population of 170 mrem/year would be 16,000 to 32,000 extra cancer and leukaemia cases in the USA alone. They therefore demanded an immediate reduction of the permissible dose to one tenth, that is to 17 mrem. Furthermore, there is no great discrepancy between their evidence and that of the ICRP, which Gofman and

[1] John W. Gofman and Arthur R. Tamplin are radiologists at the Department of Bio-medicine in the Lawrence Radiation Laboratory in Livermore, University of Berkeley, California.

121

Tamplin themselves then used to calculate that some 11,000 to 18,000 victims could be expected. We shall be returning to these two research workers later.

All these unexpected manifestations, combined with the increasing incorporation of radionuclides, dealt a severe blow to the theory of the critical organ. This theory, which is still in use today, has the advantage of being very simple to apply in practice, but conceals the serious error of overlooking the risk of cancer in total body irradiation, except in the case of the bone marrow, even though almost all the other organs are just as susceptible to cancer—the ICRP still attributes no threshold dose to most of them. The crucial source of the error lies in the underestimation of the relative sensitivity of all the tissues.

The ICRP describes this situation as follows (67): "It would be generally agreed that when the whole of the body is uniformly irradiated, so that the sensitivity of the tissue dose is uniform, the overall risk of cancer will be some sort of sum of the individual risk for each organ. Yet, at present, the dose limit for exposure of the whole body is determined by the tissue dose in the bone marrow and gonads, and no account is taken of the cancer risk from the simultaneous exposure of the rest of the body.

"The concept of the 'critical organ' does not take account of the total risk from exposure for all the parts of the body, weighted according to their relative radio-sensitivity. It is surely necessary to consider the total risk if a proper assessment of the dangers of any kind of radiation exposure is to be made.

"The fact should not be overlooked that the dose to the critical organ from any particular kind of radiation exposure does not define the overall cancer risk, which will always be greater than this—sometimes to a trivial degree, sometimes to a much greater degree."

The ICRP itself makes it quite clear here that serious

122

calculation of the risk of delayed damage is quite impossible at the present time!

But not a word of all this is breathed to the public. New bases for the establishment of radiological protection measures are discussed in a recent ICRP publication. According to this, the total risk should no longer depend on a single critical organ but rather on a summation of the individual risk to twenty-seven exactly defined bodily parts, taking into account relative radio-sensitivity (71). However these new and improved radiological protection measures cannot be implemented because the areas of radiobiology which they involve still lie largely in the dark at present (60).

We have now reached a situation where it has been recognized that the foundations on which radiological protection laws are built are inadequate. In the same way as we deceived ourselves decades ago by thinking there was a single threshold dose—as there is for many poisons —and developed the theory of the critical organ from this, so, recently, we have been making similar mistakes about radioactivity, with far-reaching consequences.

Obviously science cannot be taken to task and told it must collect its evidence step by step. Nonetheless it is irresponsible to exploit inadequate scientific evidence technologically at the cost of endangering the public. With respect to the lack of knowledge in the field of radiobiology, it must really be asked where those responsible find the courage and conviction to force atomic power stations on the public when their total radioactive effects cannot be properly assessed at present.

The authorities and those with interests in atomic power stations are appealing to radiological protection laws which are built on incomplete evidence. Protection laws should protect the whole population from disease, and not permit thousands of sacrifices to be made— otherwise they become "disease laws". *It would be unthinkable for the state to tolerate substances in food-stuffs and medicaments which have been proved to cause cancer and which will inevitably claim a certain number of victims.* But the contamination from nuclear power stations is allowed to continue. Expensive security and control measures cannot prevent it; they only serve to throw more sand in the eyes of the innocent. Ultimately control and supervisory measures which are based on ill-founded limits can only lead to a false sense of security.

As the ICRP itself writes, the establishment of limits for the population is really only a question of keeping the harmful effects to "an acceptable limit" (see p. 92). Threshold doses which carry no harmful effects have therefore been abandoned. If maximum permissible doses are mentioned today we can therefore be sure that the establishment of a permitted level of radiation for the population in question will inevitably entail a number of sacrifices. In order to be able to determine the number of victims and to establish the MPD from that number, calculations of risk must be made.

7·IMPORTANT FACTORS IN THE ESTABLISHMENT OF THE RADIOLOGICAL PROTECTION LAWS

Precise scientific data are required for making calculations of risk. The following elements are crucial and will be briefly discussed. The result will be very depressing!

Nature of disease— $\begin{cases} \text{somatic} \\ \text{genetic} \end{cases}$ $\begin{cases} \text{visible deformities} \\ \text{functional disturbances.} \end{cases}$

measurement of damage
dose size
spatial dose distribution
dose-response curve
critical organ and relative radio sensitivity.

124

(a) Nature of disease

The prerequisite naturally is that the diseases be known. We make a distinction between somatic and genetic damage (cf p. 89). All forms of delayed damage are non-specific illnesses which could also occur for reasons other than radiation. Their incidence rate, however, is raised by energizing radiation. They consist chiefly of : malignant new growths, like cancer, leukaemia, bone marrow tumours; physical and mental damage to the embryo; reduced fertility; accelerated senility; reduced resistance to infections; reduced vitality. Leukaemia and cancer have received the most attention from medical research but there are still vast areas where knowledge of these diseases is very inadequate. How all these diseases arise in relation to one another is still largely a mystery.

The extent of our ignorance of illness arising from genetic damage is highly dangerous. In that case, all kinds of functional defects can also be expected. The Austrian Medical Association, in its memorandum about the first atomic power station in Austria, writes (88): "As the formation of enzymes that link metabolisms also derives from the gene, damage to the latter can also bring about disturbances in enzyme synthesis and therefore give rise to metabolic defects. Even today we know of over 100 very severe hereditary diseases of this kind which affect the formation of the blood, the protein metabolism, the carbohydrate and fat metabolisms, the glycogen and purine metabolisms, etc, and lead to blood diseases, imbecility, epilepsy, brain damage, eczema, arthritis, skin carcinomae, damage to the retina and lenses, blinding, calcination of the kidneys, kidney stones, cramp conditions and early death, sometimes even in childhood. Others cause liver damage, shrinking of the liver or tumours, muscle weakening, defects in bone development. . . . All these diseases are incurable because they

125

are rooted in the genotype and most of them will be passed on to descendants."

Most common, naturally, are the physical deformities, yet they only represent a very small proportion of the total genetic damage. As genetic damage does not reveal itself until later generations, the number of individual illnesses cannot be predicted, for the future cannot be calculated statistically. So the damage to future generations can scarcely be imagined. Figures for mutations alone express too little. The ICRP thinks (56): "The genetic risks of radiation, especially from small doses, are much more difficult to assess than somatic risks—and will manifest themselves only in future generations."

No radiation damage to future generations can be described today except through the so-called "genetic death." This is accompanied by the elimination of one gene in a people. However this elimination of one gene tells us too little. The ICRP confirms (56): "The proportions of gene eliminations associated with the different types of damage are quite unknown and the corresponding disadvantage to a community cannot be assessed."

The ICRP writes further (57): "Because the nature of the somatic and genetic effects of radiation differ so markedly, as does the period within which they may show themselves, the numerical estimates of the risks of the two types contained in this report are a most imperfect guide to their relative significance from the viewpoint of human welfare."

From what has been said it can be concluded that nothing is known at the present time either of the kinds of damage, or of the quantities in which it will appear, or of its influence on future generations. The discharge of proven mutagenous substances—like artificial radiation —into our environment must therefore be condemned as totally irresponsible.

126

(b) The measurement of damage

For the measurement of total radiation damage, there would have to be a standard measure for both health and genetic damage. It would then be possible to make a total estimate of the component risks. The present theory of the critical tissue, however, does not allow for such a total to be calculated, apart from which there is no uniform measure at our disposal. The total risk therefore cannot be estimated at all.

The new structure of radiological protection laws discussed by the task group of the ICRP would authorize such a total of component risks to be calculated but it still avoids the fact that crucial elements (particularly the relative radio-sensitivity of the various human tissues) remain unknown and that no standard measure of damage is available.

(c) Dose size

Even this can only be estimated very imprecisely—particularly with incorporated radionuclides. Moreover it is too long since any work was done on the exact physical measurement of radiation, because the biological effects are not understood. However, we should not let this deceive us into thinking that it is possible to measure radiation physically in any case.

(d) Spatial dose distribution

This factor is frequently unknown and cannot be exactly defined. (In particular, incorporated radionuclides can be most unevenly distributed). The ICRP writes (59): "Although in some respects such a scheme would achieve a useful simplification, it might introduce some administrative complications, particularly since the spatial distribution of dose in an exposed individual is often so poorly defined."

127

This important factor in the calculation of risk is therefore generally very uncertain.

(e) The dose response curve

Attention has already been paid to this basic element. The curves for the various illnesses are still largely unknown. Obviously certain quantitative correlations with high doses of radiation are known but even the best researched leukaemia risk (irradiation of the bone marrow causes leukaemia) is still rather an uncertain quantity. The ICRP states (59): "Our basic ignorance of the relative susceptibilities of the various parts of the body to tumour induction by radiation must not be underestimated. Even for the bone marrow, the risk estimates applicable to radiation protection conditions involve a considerable extrapolation downwards in dose, and the populations from which the estimates were derived (Japanese survivors and treated spondylitics) were by no means uniformly irradiated."

At the same time this statement makes clear our ignorance about the relative radio-sensitivity of the various body tissues, which, as the next most important factor, is connected with the critical organ.

(f) The critical organ and relative radio-sensitivity

The critical organ must be known for every illness that results from radiation damage. In many cases today it *is* known. Still, the relative radio-sensitivity of the various critical organs ought to be known too. It is hardly ever the case that only a single organ is irradiated: there are always several involved, or even all. In this field nothing whatsoever is known. The ICRP thinks (70) that all the present assumptions about the relative radio-sensitivity of the various organs are provisional, because the time for which irradiated people have been observed would be too short to allow even a rough estimate of all the tissues, which have been given much less attention than the bone marrow.

At another point the ICRP writes (68): "Therefore there must be large statistical uncertainties involved in calculating the apparent sensitivity of any other tissue or organ as compared with the bone marrow. Another ten, or preferably twenty years of observation should reduce these uncertainties, but only to a limited extent."

The ICRP furthermore confirms the limited value of the present conception of the critical organ for the establishment of regulations for unequal irradiation (60). It explains that it is not yet possible to change the scheme and establish a reasonable norm for the estimation of risk. However it does suggest that the development of a new scheme should be kept in mind if new information is forthcoming. And it states quite clearly (60): "One point which emerges from the analysis is the need for a decision to be reached on the basic standard of radiation protection—whether it refers to the whole body uniformly irradiated or to the bone marrow."

8 · FACTORS NOT TAKEN INTO ACCOUNT BY THE RADIOLOGICAL PROTECTION LAWS

After all these arguments, it should have become evident even to the most hardened sceptic, that our present radiological protection laws are built on very weak foundations. But that is not all. A whole series of factors are not taken into account at all in the protection laws. The following, often extremely complex problems are involved.

(a) Individual risks (standard human being)

Radiological protection laws have been laid down without taking account of the risks to individuals. Assumed doses are simply prescribed for a fictitious standard human being. It has been impossible to concentrate on the

129

sources of incorporation because the body load resulting from incorporated radionuclides can be estimated only with great cost and difficulty. Maximum permissible doses are imposed for the radionuclides in drinking water, air and foodstuffs. According to these it should be impossible for a human being to exceed the highest safety level of incorporated nuclides. This procedure depends on the existence of a standard human being to whom the set doses are geared.

This standard human being is only allowed to drink a certain amount of water, breathe a certain amount of air and ingest certain quantities of foodstuffs—within a certain period of time. His internal organs are of a standard size and he excretes such and such an amount of urine, sweat and excrement every day. He will also have been X-rayed a certain number of times for diagnostic purposes, etc.

Unfortunately there are a lot of people who do not conform to this standard. On the contrary, they suffer in their lives quite varied radiation loads depending on the area in which they live (varying natural background radiation), on radio-therapeutic and radio-diagnostic irradiations, on the frequency of air travel and televiewing, and particularly on individual eating habits. It is therefore obvious that the establishment of maximum permissible doses for whole population groups is senseless.

If we stop to consider that a doctor has to tailor medicaments individually to each patient in order to eliminate potential side-effects because of individual sensitivity, it is quite incomprehensible to think that additional doses of radioactivity may be prescribed en masse for the whole of mankind. Moreover, radioactivity, in contrast to medicaments, can only have a detrimental effect (apart from the treatment of cancer) and it is quite impossible to take the individual into account by using the standard human being as a norm for calculations.

130

In law the doctor has no means of compelling his patient to use a particular substance with a particular therapeutic effect, for the simple reason that undesirable side-effects might set in. As far as radioactivity is concerned, the whole of society is being prescribed artificial radionuclides—which have been proven dangerous—without anybody being in a position to opt out. Woe to those who deviate from the standard—and to all the others too!

(b) Age

The radiological protection laws for the peaceful application of nuclear energy also take age too little into account. The adult is much more resistant to many radiological diseases than the adolescent or the child; the embryo in the mother's womb is the most sensitive. However, our planet (including the areas around nuclear power stations) is populated by a colourful cross-section of people of all different ages. There are no two standard people walking about. Far too little attention is paid to this fact.

At the information session of the Swiss Atomic Energy Commission, Professor Fritz-Niggli stated that there was no such thing as a radiological threat to the embryo in the vicinity of nuclear power stations—although the growing being was more sensitive to radiation than was normally the case—because short-term irradiation of 5 rad would be required before physical deformities arose (33). Such statements must make the layman think that everything is fine and that radiobiology can produce perfect, definitive judgements.

However, earlier research had already led us to suspect that the embryo in the first three months of its development could be much more sensitive to radiation than the adult (105, 109). Injury does not necessarily always show itself through physical deformity. The radiological protection laws in operation today were set up back in the fifties

131

and are based for the most part on the lesser sensitivity of the adult. The significance of the age factor was not understood at that time.

In spite of this, protection practice often shies away from accepting new scientific evidence if it does not happen to suit those who promote the peaceful application of nuclear energy. Max Planck is once supposed to have said that it often took decades before new scientific evidence got through—in fact until most of the professors of the old school had died off, and most of the students too! Given the rapid changes that are taking place in everything nowadays, it should not take nearly as long, yet aspects of nuclear theory are allowed to stand without being corrected.

Back in June 1970, Dr A. Stewart of the University of Oxford pronounced that the human embryo could be some 500 times more susceptible to leukaemia and cancer than the adult (109). The research published by Professor Sternglass at the end of 1970 and in mid-1971 pointed in the same direction. It confirmed that child mortality increased in the vicinity of nuclear reactors. According to this research, for example, the boiling water reactor "Dresden" (180MW) in Chicago, within a ten year period of operation from 1958 to 1968, caused the deaths of 2,500 children in the area down-wind of the exhaust plume (105). Similar observations could be made for other reactors, and Sternglass also had his statistical calculations checked by an independent source (24). It sounds very unlikely, as a Swiss authority has maintained, that Sternglass could have made a mistake with his wind directions (92). Professor Dr Werner Zimmermann, Ringgenberg, notes with full justification (146): "The objection was raised against the statistics of Mrs Mary Weik, Box 148, 150 Christopher, New York 10014, which came up with similar results for leukaemia and cancer, that she was 'only a housewife' and her work was therefore valueless. This cannot be said of

132

Professor Sternglass (Director of Radiological Physics, School of Medicine, University of Pittsburgh) and cannot be shrugged off with fantasies about wind directions."

Professor Sternglass is supposed to have demanded the closure of all boiling water reactors and a lowering of the radiological protection limits (29). But even these measures would take no account of the potential biological concentration of radioactive cesium, strontium and yttrium, sister products of the noble gases discharged from the reactors.

(c) Constitution, condition and individual sensitivity

Because the radiological protection laws are based on the average person, not enough attention can be paid to individual risks like constitution and condition. Dr Herbst says of this (140): "The individual conditions which result from the action of energizing radiation on an organism also decide whether the person in question becomes ill or remains healthy. The same irradiation of different organs also causes very appreciable differences in the reactions as well as in the effects on health. These differences are directly related to the state of development, constitution and condition of the individual."

Here again we can see the inhumanity of using the mass-prescription of increased levels of radiation of the kind produced by nuclear power stations.

Conditions can be adversely affected even by a slightly increased environmental radioactivity level. Dr C. E. Mehring at the International Convention for Vital Substances in Montreux in 1971, reported his extensive research in Germany, which pointed to a correlation between a slight increase in the environmental radioactivity level (from atom bomb tests) and a worsening of the condition of the population. He could prove statistically an increase in some diseases (e.g. tuberculosis, perforation of the appendix) as a consequence of two periods of increased environmental radioactivity in the fifties and sixties, at which time the younger generation was particularly badly affected. He fears that a general increase in commonplace illnesses could arise as a result of a radioactive contamination of the biosphere.

With this new risk factor, which consists of a reduced body resistance as a result of the removal of leucocytes, the ionizing radiation has no effect in the sense of delayed damage, but instead paves the way for many other illnesses. Small permanent doses of artificial radioactivity are therefore most insidious because the source of illness cannot be traced but can only be understood in terms of statistical correlations.

Besides this, it should not be overlooked—as one of the delegates at the Vital Substances Conference stated—that other members of our ecosystem could also suffer a more or less similar deterioration and are also exposed to the danger of ruthless destruction. The damage which can be done through such artificial attacks on nature is quite unimaginable.

(d) Nutrition
Nutrition is also important and is itself subject to so many variations that effects cannot be calculated with standard measures. According to Herbst (140), there is a growing body of evidence to show that nutrition patterns correlate with the reaction of the individual to energizing radiation and, consequently, with the effects on health. Given a total body load of 450 Roentgen units, if one half of those affected die and the other half become very ill but remain alive, the difference in reaction to radiation depends not only on constitution and condition but also to a certain degree on nutrition.

For example, research workers have established with rats (140) that recovery and survival after high doses of radiation can be achieved by feeding with proteins.

134

Similarly lipoids and essential fatty acids also lower the mortality rate after acute radiation injury. Moreover, the calorie content of food also appears to be influential, although fatty food has detrimental effects on irradiated rats. Injections with a mixture of various vitamins and β carotene proved relatively resistant. Even the condition of the intestinal tract can play a part. It seems, for example, that plutonium is better absorbed, the more acidic is the intestinal area.

Practical experience proves that, particularly in the lower levels of radioactive contamination of the nutritional environment, it is not external radiation but the incorporation of radioactive materials that is the critical factor. It is estimated that 95 per cent of the radioactive atoms in the body found their way there through food. In general, the radioactive isotope of an element is the better absorbed, the greater the deficiency of that element in the organism. So one-year-old children suffering from iron deficiency absorbed up to 75 per cent of the iron 59 from the intestinal tract. Children with normal amounts of iron only absorb 5 per cent.

(e) Synergic effects

Radioactivity also has the property of heightening certain poisonous effects. Various poisons or harmful substances (noxa) not only accumulate but, in combination with energizing radiation, also produce effects which are greater than the sum of the individual noxa. This synergic effect of radioactivity can also arise with many chemicals, through physiological factors, infections, malnutrition, temperature impulses, etc. Dr Herbst says (44): "Energizing radiation can sensitize the effect of other noxa even in the smallest doses." It is therefore hardly surprising if it is proved in many experiments with animals that energizing radiation not only causes increases in the mortality rates from cancer, but also in many, if not quite all the remain-

135

ing deadly diseases. None of these additional injuries are taken into account by the radiological protection laws.

(f) Chemical genetic damage through poisons

It is known that chemical poisons can also have mutagenous effects. None of the calculations of risk included in the radiological protection laws take the slightest account of eventual, hereditary damage through other harmful environmental agents. Because of the world-wide dissemination of the poisons civilization has developed, primarily pesticides, such considerations have become impossible in any case. Radioactive damage to the genotype will in fact combine with damage already caused by the effects of poisons and accumulate. The total damage is inconceivable. The statement that the natural mutation rate will be increased by such and such an amount per 1 rad can lead to no conclusions about the total changes involved. In any case, figures for mutations tell us too little to allow us to make an estimation of the dangers that threaten us through nuclear fission.

(g) Transmutations

These evidences of radioactivity, which can give rise to even greater biological effects than energizing radiation itself, have already been described on p. 109. The radiological protection laws take not the least notice of these potential sources of harm.

The dropping of the atomic bomb has certainly provided science with a great deal of evidence about energizing radiation. However, the kinds of damage described here, which are not taken into consideration by the radiological protection laws, show that even today there is an almost total lack of experience of the true biological effects of continuous exposure to chemical and radioactive poisons.

The present radiological protection laws are just patch-

136

work over hollow ground. They do not entitle anybody to compel whole populations and entire societies to accept – fictitious permissible doses of radiation.

9 · MISLEADING THE PUBLIC

The foregoing discussion has brought up several concrete examples of the ways and means that are being used to misinform a gullible public on the dangers of atomic power stations. The discerning reader should now have sufficient well-founded evidence at his disposal to be able to judge for himself. The following model illustration will allow us to examine quite soberly and coldly the irresponsibility of maintaining the maximum permissible doses which were valid in 1971 for the peaceful application of nuclear energy.

After a two day information session held by the Swiss Atomic Energy Commission in Berne in 1971, there was a large press conference. According to the *Tages Anzeiger* (*Daily Advertiser*) of 18 November 1971, Professor Dr Fritz-Niggli, Director of the Institute of Radiobiology at the University of Zürich, assured the journalists (141), that the maximum permissible dose (500 mrem/year), established by an international panel of radiobiological experts, both for domestic use, and in the vicinity of nuclear power stations, "could be *in no way dangerous* either *directly* through damage to health or *indirectly* through damage to the genotype, either to those in the vicinity of a nuclear reactor or to mankind as a whole."

Any schoolboy can work out the following equation: according to the statements of the ICRP, it can be assumed that cancer results 5–6 times more frequently from irradiation than leukaemia (64, 80). As a result of the most extensive research on leukaemia, it can be assumed that, according to degree, each rem of additional radiation load per 1 million people will give rise to about 20 cases of

137

leukaemia in 10–20 years (55) and also, therefore— because of the longer latency period of about 30 years—to 100 (5 × 20) cases of cancer, which gives a total of 120 cases of cancer and leukaemia (in 30 years).

With the additional load of 500 mrem/year permitted by radiological protection legislation, and in fact exhausted in the course of a life span of sixty years, a total of 30 rem (500 mrem × 60 years = 30,000 mrem = 30 rem) would be absorbed. This gives us cause to fear that, if 120 cases of cancer and leukaemia result from an irradiation of 1 rem on 1 million people in 30 years, a dose of 30 rem would cause 3,600 sixty-year-old victims (30 rem × 120 persons). The probability of illness for this age group therefore amounts to some 1 : 300 (1 million : 3,600 gives about 300).[1]

In this way we could calculate that the probability of illness would be about 1 : 1,000 for the maximum permissible additional load for the peaceful application of nuclear energy (namely 170 mrem/year).

As has already been mentioned, the Swiss Atomic Energy Commission maintained in a full-page article in the *Neuen Zürcher Zeitung* of 26 August 1970, that the recommendations laid down by the ICRP could not give rise to "any kind of detrimental effects either to health or heredity."

Such flagrant misinformation of the public requires factual explanation. Such practices mean that the good faith of the public is openly being scorned, not only by the atomic industry and the authorities responsible for the provision of energy, but also by the scientists. However, there can be no doubt that the efforts made by radio-

[1] In this calculation, the latency factor was excluded in order not to complicate the matter. In principle, this does not matter because the basic facts are imprecise. The damage would doubtless be much greater because, for example, other diseases and the genetic effects have not so far been taken into account.

138

logical protection research and the opponents of atomic energy will ultimately bear fruit, that the myth of the peaceful atom for everybody will be exploded and come to sound more and more improbable, and that the maximum permissible doses will have to be massively reduced. Then, no doubt, the nuclear industry will explain that high doses are no longer necessary anyhow, as a result of more advanced technology. But who will believe that, after the unfair practices of the past? Even now it is being suggested that it may be possible to keep back all the radioactivity in the reactor. Not even the atom experts believe that. Professor E. Tsivoglou, who was commissioned by the canton of Basel to work out calculations for the atomic power station, Kaiseraugst (121) writes: "In the view of the author 'zero' represents an ideal limit which, although desirable in principle, cannot in most cases be achieved in practice."

The public cannot control discharges from the operation of nuclear power stations. We see nothing, hear nothing and smell nothing and society would do well, particularly after previous experiences with industrial pollution, to ban such processes entirely. They are insidious, deadly and outside its control. They should be strictly forbidden—even if it one day became possible to reduce discharge to an absolute minimum, or even to nothing.

The application of nuclear fission has been a scientific and technological blind alley and will remain so; it must be abandoned altogether, irrespective of safety measures. The potential danger at the present time is of such monstrous proportions that it is almost unimaginable.

10 · ACCIDENTS

The weakest and most potentially dangerous part of a nuclear power station (in light water reactors) is the reactor pressure vessel. It is usually made largely of steel

139

and cracks can appear as a result of the destructive effect of the neutron rays—for radiation is not only capable of strengthening metals, it can cause them to become brittle at the same time. Added to this, cracks cannot be recognized early enough because the vessel cannot be inspected when it is in operation. Moreover, for economic reasons, reactors are being built to be more and more productive and this requires larger reactor vessels. In consequence the chances of an accident are higher.

The word "accident" can only be used in a restricted sense for nuclear technology, however, because there will never be many spectacular accidents with lots of dead or injured. Instead there are increasing numbers of incidents whose only distinguishing feature is their complete lack of effect on our sense organs. The results only become evident when radioactive substances have escaped and can circulate freely and effectively.

When this occurs, or in the case of an explosion in the reactor (more delicately termed an "excursion" for innocent ears), then radionuclides can be carried away by air currents across individual countries and whole continents. A part of this discharge then seeps into the ground or is washed into the waterways. This does not mean that the active substances have disappeared: they remain active for years, decades and generations, according to their half-life.

With nuclear power stations we therefore have a quite different problem on our hands from the usual ones of accidents in factories, in traffic, in aeroplanes, in explosions in dynamite factories, in dam bursts, etc. Comparisons between the radiation risk from a nuclear reactor and such accident statistics are therefore misleading for they only serve to exploit the ignorance of the public.

Although in a nuclear power station "accident" people are not necessarily killed or injured, those involved can still die years later from delayed damage, resulting from

140

an increased radiation dose or from the incorporation of discharged radionuclides. They are naturally not included as victims in accident statistics. A human being can stand up to a 25 rem dose at one irradiation without any consequences becoming immediately apparent clinically, because doctors are in no position to diagnose radiation illness. The biological measurement of dose is not yet that far advanced. Eventual genetic damage is particularly difficult to detect.

The damage that an atomic accident causes is in no way over and done with with the particular incident but continues to be effective into the unforseeable future. Comparisons with other risks inherent in human civilization which are based on numbers of dead and injured are therefore senseless—they only serve to dodge the issue. The development and practice of nuclear technology is riddled with such incidents and accidents. We have a very clear account of the period from 1944–1964, because an exact record of every nuclear incident is available in a book by Dr Erich Schulz (101). This makes it quite clear that 900 such incidents or accidents occurred in 15 countries in twenty years—among them 24 cases of death through criticality,[1] contamination and radiation. Some 1,800 people received increased radiation doses, although naturally they do not figure in any accident statistics. Excluding military accidents there still remain 845 incidents with 18 dead and 1,400 people with increased radiation doses. It is significant that 67 per cent of these incidents can be traced back to human error.

A classic example of a potentially dangerous episode is provided by Windscale in England, where 20,000 curies of iodine 131 escaped into the surrounding area in 1957, along with strontium 90 and 89, cerium 137 and tellurium

[1] Condition in which a self-perpetuating chain occurs in the reactor. If disturbed, explosions (so-called excursions) could occur.

141

132(101). Radionuclides were carried by wind currents to Western Europe where an increase in background radioactivity was confirmed. In one area of 300 square miles, all the milk had to be confiscated for two months. Plants and vegetables were contaminated, but the authorities decided that vegetable nutrition only played a very small part in the population's diet and the medical advisory council authoritatively declared that no damage to health could result from the accident, either to the staff or to the public.

Week by week and month by month, reactor faults and accidents are, where possible, hushed up. There is unfortunately no record of reactor accidents for the years 1964–70. Nevertheless the *Centre d'Etudes des Risques Atomiques* in Brussels, which has to be informed of all radiological accidents, has supposedly been told of several hundred such incidents, which have unfortunately remained largely unpublicized (6).

In 1969 the Swiss atomic reactor Lucens had to be stopped after only a few hours in operation because of a fault. There is supposed to have been no substantial contamination of the vicinity because the reactor was built underground in rock. On the other hand, extensive contamination occurred in Rocky Flats, USA, where a plutonium plant was destroyed by fire in the same year. Nobody knew anything about it until Congress was asked for credit of 45 million dollars to have it rebuilt.

But normal commercial power stations are not 100 per cent free of faults either. In August 1969, the River Ems in Germany was reportedly contaminated with radioactivity for two days as a result of defects in the nuclear power station, Lingen(6). Naturally nobody went down in the accident statistics as being dead or injured. Nevertheless the radionuclides discharged can harm people who may possibly only die years later from some form of delayed damage. At that time there will be no way of proving

142

whether the illness is a so-called spontaneous one which has been contracted from causes other than energizing radiation.

How the public is informed of these matters was the subject of a lecture given to the information session of the Swiss Atomic Energy Commission in Berne, in which it was stated (41): "In all seriousness and good faith it can be stated that, at least in the whole of the Western World, not a single person has yet received injury through radiation, either in the immediate or more distant vicinity of nuclear plants." Truly a courageous statement!

A real catastrophe can occur with the so-called GAA (greatest acceptable accident). However, a reactor does not explode like an atom bomb but more like a steam kettle. The safety vessel is constructed in such a way as to allow for such an "explosion" in the primary cooling system but there can still be no absolute guarantee of safety. The fissile contents of commercial power stations are in fact several times greater than those of an atom bomb. According to Professor Dr Bechert, the contamination effect of an accident at the Wurgassen nuclear power station in Germany, for example, from which only one per cent of the radioactivity contained could escape, would correspond to the detonation of a medium sized atomic bomb(6).

If such a release of radioactivity were to occur in the reactor then, according to Dr Groos, chief advisor for the security of nuclear reactors in the German Ministry of Science and Research (95), "a whole chain of functions is set in motion to ensure that nothing escapes into the outside world. But if there is some trivial fault—a valve not quite shut or something else not functioning—then the GAA can look quite different."

"Fate pays no attention to the word impossible. We can, for example, prove cases in which a mathematical impossibility has taken place." So said Dr David Okrent, a

143

former chairman of the USAEC advisory committee, before Congress in 1967. Dr Adler, President of the Swiss Federal Commission for the security of atomic plants, has also stressed the frequency of accidents in the so-called "Farmer-concept", according to which the risk of accident (with simultaneous release of radioactivity) is so fixed, that "such an incident is, on average, only to be expected once every 1000 years" (2). In the foreseeable future, however, there will be a thousand power stations in operation, so that an average of one accident per year could be expected. This calculation of risk demonstrates the dangers of accidents in nuclear power stations, which, as long as no steps are taken against them, are constantly expanding.

It was not without reason that the German Federal Minister, Leussink, held back authorization for two years on a scheduled BASF nuclear power station in a connurbation area of Ludwigshafen and on a sodium cooled reactor (prototype) near to Aachen. The President of the German Department of Education and Science made the following statement on 19 August 1970 (90) : "Both decisions are based on the possibility of faults in nuclear power stations. The probability of extreme malfunctioning with the release of radioactive nuclear products into the environment is indeed very small; however, the effects of such malfunctioning can be very great. Although great efforts are undertaken to minimize the possibility of breakdowns and their effects through expensive, systematically planned, technological safety measures, it is impossible to achieve absolute security with nuclear power stations as with other complicated technological equipment; there remains a risk, either of human error, or of inadequate experience, or of statistical sources of error. In order to try to reduce this risk as far as possible, it has been policy the world over never to build even well protected reactors in the vicinity of towns."

144

In addition to this, the city of Mannheim, a group of 50 doctors, and the Weltbund zum Schutze des Lebens, as well as individual people had already objected to the scheduled BASF reactor.

In fact this case shows quite clearly how the building of nuclear power stations is carried on under the cloak of a moral double standard. We are prepared to expose thousands or tens of thousands of people to risk in the less densely populated areas where reactors are built but (fortunately) still exercise constraints for the hundreds of thousands of potential victims in the vicinity of large towns.

The somatic population doses are generally included in the calculations for the choice of a site for a nuclear power station. These doses are always possible as a result of radioactive contamination of the environment during normal operation, of malfunctioning and, of course, of extraordinary catastrophes. Dr Herbst's opinion is this (44): "Apart from the fact that these methods take no account either of the risk of delayed damage or of ecological processes, it should also be noted that those who will ultimately be affected are bundled together without reference to the highly individual risks involved, and assessed according to unacceptable mass-theory, solely in terms of the numbers of people affected or of short-term deaths which result from malfunctionings or catastrophic accidents."

The so-called Brookhaven Report by the USAEC (American Atomic Energy Commission) dating back to 1957 has become famous (22). In this it is estimated that the explosion of a 100-200 megawatt reactor in a lightly populated area would cause the immediate deaths of 3,100 people and injure a further 43,000 (some of them with incurable radiation diseases). An area of 15,000 to 230,000 square miles would be contaminated: more than the area of Pennsylvania, New Jersey and New York put

145

together! Significantly the report was withdrawn after a short time.

In the summer of 1971, the American atom experts Jan T. Forbes, H. W. Kendall, J. J. Mackenzie and D. F. Ford are reported to have confirmed, in a sensational study, that the present emergency cooling systems in reactors could fail. In the event of this happening, the reactor nucleus could burn through the concrete foundations and the external containment of the reactor and release a gigantic cloud of radioactive fission products. The consequences would obviously be quite horrific.

This study is supposedly based on experiments carried out on a small scale, and the experts are still arguing whether such results could be validly transferred to large scale production reactors. Obviously a 500-1000 megawatt reactor could not be blown sky high just to settle the matter. However, the recently discovered doubts about the emergency cooling system could be most alarming because tens of thousands of people could be forced to live under this dreadful threat. (The author passes on this American information with the reservation that, at the time of writing, he did not have full information about it at his disposal.)

Apart from the risks of accident to reactors, the dangers which may result from the changing of used reactor rods and from the transportation of old fuel elements to the reprocessing plants should not be underestimated. These elements contain considerable quantities of radioactivity and probably represent the most dangerous freight ever transported by man. The possibilty of an accident hardly bears thinking about. Packaging the elements in containers cannot provide total security.

The potential dangers from the nuclear power industry have no equivalent and accidents can never be entirely eliminated.

146

11 · RECENT RADIOBIOLOGICAL RESEARCH

(a) John W. Gofman and Arthur R. Tamplin

One cannot discuss the dangers of nuclear power stations today without mentioning these two American scientists. The publication of their research and their explanations of it in Congress were the most important events in this field in 1971 (36, 37, 39). Dr Gofman is Professor of Medical Physics and Dr Tamplin works in bio-medical research: they are both at the University of Berkeley in California.

Their work showed that the 170 mrem/year additional dose allowed by the Federal Radiation Bureau for the peaceful application of nuclear energy in the USA would, according to their calculations, cause 16,000 to 32,000 cancer cases as well as 150,000 to 1,500,000 extra deaths through genetically transmitted diseases. They therefore demanded a lowering of the limit to one tenth, that is to 17 mrem/year. This naturally provoked the most strenuous opposition from the American Atomic Energy Commission (USAEC), which has ties with the nuclear reactor industry, as well as from other similar bodies.

Gofman and Tamplin also drew the public's attention to the untenable position of the USAEC which, on the one hand, has the legal obligation of furthering "the development of the peaceful atom" and, on the other hand, is supposed to protect the health and safety of the US citizen. Such a dual responsibility should never be carried by one body alone, because the promotion of atomic energy runs quite contrary to any interest in the protection of the population from the dangers of radioactivity. One critic compared it with having the chicken run guarded by the wolf!

It seems just as grotesque to think that Gofman and Tamplin, or the Lawrence Radiation Laboratory, have carried out their research at the request and with the support of the USAEC. They began in 1963 with a brief requiring them to examine the dangers from the release of radioactivity both for the biosphere and for human beings.

Fortunately they had the courage at that time not simply to come up with a policy verdict, which would certainly have suited the USAEC, but to publish their results quite frankly and openly. The USAEC then showed its true face and acted according to the usual recipe of the atomic industry: that of discounting businessmen, scientists, experts and even Nobel Prize winners, if they do not happen to share USAEC opinion.

What is quite terrifying about this situation is the fact that the atomic industry as a whole follows unquestioningly in the wake of the USAEC. The Swiss Atomic Energy Commission, for example, at its information session in Berne in 1971, alluded to the controversy caused by Gofman and Tamplin in the USA but did not draw any conclusions or show itself in the least impressed. In essence its comments are embodied in the following (significant) words (102): " 'Dr Gofman is wrong by a very large factor' is the USAEC's answer. Twenty-nine leading scientists have supported it in an open letter. . . . No end or respite to the controversy is to be foreseen in the USA. . . . The most stubborn opponents of nuclear energy will presumably be convinced ultimately, after the numerous nuclear power stations being built in the USA have been operating safely for a number of years."

Such a credulous, uncritical position is more than a little suspicious. However, Gofman and Tamplin do not stand alone. No lesser person than Linus Pauling, Nobel Prize winner for chemistry in 1954 and holder of the 1963 Peace Prize, confirms their findings (89).

Amongst other things, Gofman and Tamplin state: "The myth of a safe release of radiation has at last been destroyed. Each piece of evidence that we have examined

148

has led us to the firm conviction : no amounts of radiation are safe. We have the intention of working through a critical estimation of the safety of atomic energy, in such a way that it can be openly checked both by science and by the public. The objection that the public might become over-alarmed or get out of hand, we find unjustified. If our results are contrary to the ones accepted so far, then the public has a right to feel concerned and should be in a position to demand that technological progress should only be carried on in such a way that open questions of health and welfare are decided by everybody."

The whole Gofman and Tamplin controversy is a prime example of scientific evidence being ruthlessly suppressed if industrial profits are threatened.

The two scientists have published further works on this theme with the striking titles *Population Control through Nuclear Pollution* and *Poisoned Power* (39, 38).

(b) Professor E. J. Sternglass

This man's research could hardly be more overlooked today. E. J. Sternglass, Professor of Radiology and Director of Radiological Physics at the School of Medicine in the University of Pittsburg, Pennsylvania, has constantly warned against the possibility of increased child mortality as a result of radioactivity from atomic bomb tests. Most recently, he has been putting forward convincing evidence which points to increased child mortality in areas affected by the exhaust plume from nuclear power stations. Because this fact has already been confirmed positively for four separate reactors, the chance of coincidence is slight.

Sternglass writes (105) that it was not possible until a few years ago to prove directly the deleterious effects on health of low levels of radiation or of small doses, in relation to fallout from nuclear weapons or from atomic power stations. However his research subsequently led

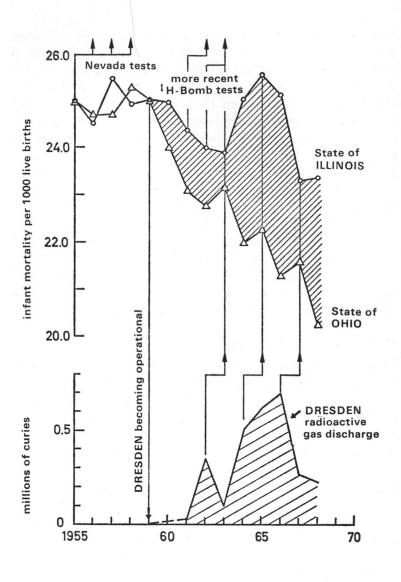

Fig. 4. Infant mortality in Ohio and Illinois, the site of the "Dresden" reactor, in the years 1955-68.

him to the conclusion that crucial damage to health, such as demonstrably increased child mortality, could be established in normal operation. The proof is provided in an analysis of the changes in death rates of children up to one year old in the vicinity of the nuclear power station "Dresden", near Morris (Illinois), 50 miles south of Chicago. The "Dresden" boiling water reactor came into operation in 1959 and has a strength of 210 megawatts. The yearly discharge of gas is known. In 1967–68 there were 240,000 to 250,000 curie and the radiation dose amounted to 13–40 μR/hour under the exhaust plume, at a distance of about half a mile; this corresponds to 114-350 mrad/year. Even at a distance of 9 miles, the plume was still detectable with 22 mrem/year. These doses are more than 10,000 times great than the 0, 001 mrem/year which the US population should receive as an average radiation load from all reactors, according to the regulations of the USAEC! The confirmed increase in child mortality can be seen from the three diagrams, figs 4, 5, 6.

In fig 4, the child mortality rate in the states of Illinois and Ohio is compared for the years 1955–68. The yearly reactor discharge is given in curies. When the reactor became operational in 1959, the child mortality rates were about the same for both states, although generally rather higher than normal as a result of atomic bomb tests. It would take too long to discuss these influences here. However, it is important to note the sharp climb (1963) and the generally higher constant level of child mortality in Illinois as opposed to Ohio, where the level fell off.

Sternglass was also able to detect a similar situation at the "Big Rock Point" nuclear power station (107). Fig. 5 represents the changes in child mortality between 1962 and 1967 in ten areas close to reactors, as well as the annual discharge of gaseous radioactive substance. Here too the influence of the reactor can be distinguished in the

151

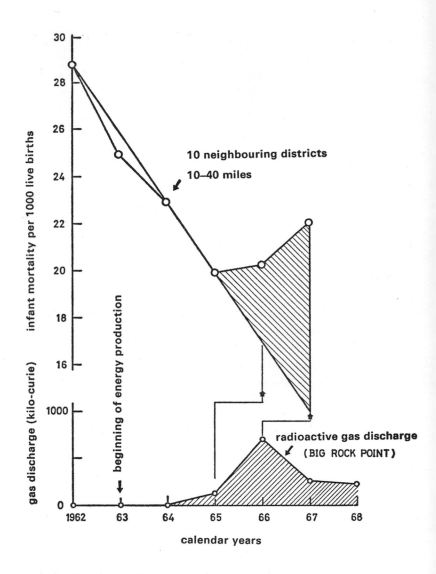

Fig. 5. Changes in infant mortality in ten districts close to the Big Rock Point Nuclear Power Station in Charlevoix, Michigan, between 1962 and 1967. The yearly discharge of radioactive gasses from fission- and activation-products is also indicated.

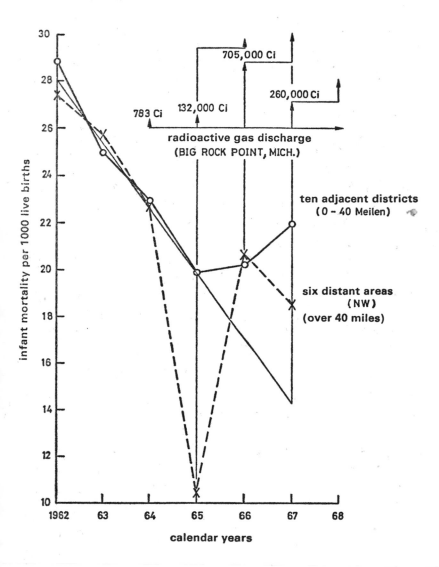

Fig. 6. Comparison of infant mortality in outlying regions in the north and northwest of the U.S.A. with rates in areas situated within a forty mile radius of the Big Rock Point reactor, Michigan.

hatched parts. Fig 6 compares the child mortality in the more distant areas to the north and north-west with that of areas situated less than 40 miles from the same reactor. From 1962–1964 both groups display almost equally falling child mortality rates. The fall in the outlying areas continued, in spite of an increase in the discharge from 783 curie to 132,000 curie/year in the years 1964–1965. However in 1966, the child mortality rate even in these distant areas increased as the discharge rose to 705,000 curie, and sank again in 1967, when only 260,000 curie were released.

It can be seen quite clearly from this that even areas 40 miles away from the reactor can be affected—during normal operation! The 705,000 curies only represent 2·27 per cent of the discharge level allowed in law.

Professor Sternglass was able to detect similar conditions in the "Peach Bottom" Nuclear Power Station in the region of York (Pennsylvania) (108). Moreover it appears from his work that those areas fed by rivers which have nuclear power stations situated on them are particularly endangered. Sternglass further states: "It seems that this gaseous discharge not only penetrates the body through the lungs. The sensitive embryo is also contaminated by the water, milk and vegetables of the area, for fission products fall back to earth. Having been able to establish from four independent sources that there is a significant increase in child mortality resulting from amounts of radioactive discharge which only represented a few per cent of the limits allowed by the AEC, it is quite clear that a radical reduction of these limits is absolutely vital for the protection of infant life."

(c) Is the ICRP waking up?

After the ICRP's 1969 *Publication 14*, in which the perilous state of radiobiology in relation to maximum permissible doses was openly clarified, things remained

154

quiet for a long while. It is well worth noting that this publication was the work of a group appointed by the ICRP to review the basic correlations. It made known that the theory of the critical organ was unsuitable for the calculation of risk, in spite of the fact that our present radiological protection legislation is based on this false premise.

According to the Swiss AEC Bulletin No 3, 1971 (21), the ICRP appointed a new task group to study the suggestions of ICRP *Publication 14* with reference to the possible modification of ICRP dose limits, in the light of present biological evidence, and to advise the commission on the practical effects of various possible systems of dose limits on radiological protection and, particularly, to present a definitive suggestion of how the present system could either be maintained or altered.

This task was undertaken in 1970 and it could take several years before any decisions are reached. Meanwhile the nuclear industry muddles quietly along in the same old way with controversial and, above all, *ill-founded radiological protection laws*. Ever newer and mightier nuclear power stations are being built, without any account being taken of ecological laws and the potential effects on the population.

This fact has been little altered by the USAEC lowering the maximum permissible dose for nuclear power stations from 500 to 5 mrem/year in the summer of 1971—a decision which was taken almost simultaneously with the, presumably not coincidental, resignation of the President, Professor G. T. Seaborg. The crucial international limit of 170 mrem/year, however, is to remain.

This new regulation has done something to help re-establish scientists like Gofman, Tamplin and Sternglass. A part of the optimistic super-structure, which has been stubbornly justified for years, suddenly fell apart and it is to be supposed that the atom industry the world over will have to stand down. *However these new limits offer no security either, because the substances which are discharged are cumulative poisons and can build up in the biomass. They are, moreover, neither biologically nor artificially destructible and biological damage begins with a radiation dose of zero.*

III · How can man survive?

As we said at the beginning of this book, the dangers of nuclear power stations cannot be understood in isolation, but only within the context of the threatened state of the whole environment. The atomic industry points to the tremendous gain in energy compared with the minimal risk of radioactivity. Two things must be said about this:

(1) The radiation risk from nuclear power stations has become a very real danger because it is impossible to estimate the total risk involved. The incidence of cancer and leukaemia only represents a fraction of all the potentially harmful effects that might ensue. The ICRP itself writes in *Publication* 9 (49) that the Commission believes the existing limits would allow a reasonable diffusion of nuclear energy in the foreseeable future. *However it must be stressed that these limits do not represent a reasonable compromise between harm and benefit, because neither the risks nor the benefits can be estimated that would justify this amount of radiation being released.*

(2) The primary question today is not: how we can use all the means at our disposal to create more and more energy; but rather: how can we survive? The official in the Ministry of Power obviously only sees the immediate task of producing more energy and never stops to ask himself whether it all makes sense.

157

In order to understand this, we must be fully aware of the fact that the progress we worship is purely material in character. Spiritually, we are back in the Dark Ages. As a consequence, we are putting into operation reckless and short-sighted methods of destroying ourselves just to achieve material benefits.

However, human beings must now realize that further purely material expansion in the fields of science and technology at the expense of the natural environment must not be allowed to continue. Man is not spirit alone he is also a quite ordinary biological being, subject to ecological laws in just the same way as all the other organisms and populations.

Science was very late—possibly even too late—in recognizing the meaning of the ecosystems which conserve life with their buffer qualities and are stabilizing forces. Thanks to these forces, life has been able to develop and the ecosystems themselves have reached a state of equilibrium. Mankind, too, is born into the security of the whole biosphere and, as a population, linked for better or for worse through ecological laws to animals, plants and the rest of the environment.

Through massive over-population and material expansion, man is now threatening to put a bomb under his own security, for today he is not only destroying single ecosystems—a forest, a bay or a river—but is endangering all the ecosystems together, even the sea. If he continues along this path, he will destroy himself.

Man as an individual possesses the faculties of perception, foresight and high intelligence, as well as free will and conscience. He is therefore totally responsible for the helpless animal and plant world and for future generations, and must protect the whole of life on earth. Only by doing this can he maintain his position.

To this end, mankind must now consciously strive to regulate and restore the ecological equilibrium, particu-

158

larly for the benefit of the individual population. Two fundamental factors have to be observed in doing this:

Birth control and family planning;
In the development of our civilization, the conservation of life and the environment must in future come before all technological and economic considerations. This must even be made law, which means putting a quite different emphasis on life styles.

Unfortunately birth control is at present facing resistance of a moral, religious and ideological nature, and public enlightenment will have to be developed to a stage where one day it will be considered immoral to oppose suitable methods of family planning.

The reverence for individual life at all costs must no longer be allowed to take precedence over the reverence for the whole of life on earth. In the framework of the whole of life, no single group—by evading the laws of life—should be allowed to disturb the ecological equilibrium. That could lead to the destruction of all life (26).

Having recognized the essential correlations, man therefore has the task of exercising controls. It would be quite wrong just to rely on God and do nothing. In that case God would not have needed to give man reason, mind and all his faculties, and a fatalistic approach to life would suffice. Man could have led his own life as he pleased, without taking account either of other people or of future generations. "Après moi le deluge"—that would be fine but, "Make of the earth your slave", does not mean that we should destroy it.

People must develop a new sense of responsibility towards the common life. Individual, material interests should no longer be given preference over the whole of life. Everyone who is able to accept the responsibilities of family, work and state today will also be able to recognize

159

and understand this new responsibility. The prerequisite for this, however, is enlightenment.

Most people today are still totally ignorant in this field, in spite of a relatively high level of intelligence and learning. Our conception of education is primarily responsible for this state of affairs. For example, in 1968 the new Swiss high school leaving certificate did not even include biology as a subject. Instead there is something called "natural history", which for centuries has limited itself to teaching systems of classification and anatomy (119).

The majority of the present high school graduates therefore know much more about cavemen and Roman or Greek history, than they do about the ecological conditions that are essential for an understanding of mankind's present situation or for thinking about the future. We urgently need a conception of education that is geared to the present and the future and which starts from the premise that nature and culture are inseparably linked (119).

The American scientist Ralph Lapp has pointed up this precarious situation with a pertinent analogy : mankind is travelling in an express train, which is racing past an unknown number of points with ever-increasing speed towards an unknown destination, in spite of the fact that there may be demons sitting on every set of points. The fearful thing about it is that in the locomotive there is not a single biologist or ecologist to be found. Almost the whole of humanity is squashed up together in the last carriage, looking backwards and teaching their children to look backwards too. This analogy involuntarily makes one think that the humourist who suggested that the only way of saving mankind was a dictatorship by biologists and ecologists, was not entirely wrong.

A great deal of expository work will be required before people will come to accept the necessary laws as quite normal and natural. This is particularly true of the long overdue laws for the conservation of the environment, which will demand sacrifices from every single individual. However, these laws should not deceive us into thinking that *only the symptoms* of the pollution and destruction of our environment can be alleviated for a certain time (like sewage farms!). They are certainly extremely important as stop-gap measures but will not simultaneously eliminate *the causes*. There will actually be no way other than to seek out the equilibrium and to check unlimited growth of every kind.

The favourite image of one's own little brood of children will have to become a thing of the past, and strict limitation to two children will become compulsory—even by law! Those who want more children will have to adopt them. This has already been suggested in a forty-four point programme for environmental protection proposed by the World Wildlife Fund. It is certainly better for mankind to regulate itself, for nature has always managed to deal fairly severely with over-population in the past. The advance of civilization has brought man such terrible powers of potential destruction that irreparable damage could be inflicted on nature until mankind's numbers were decimated and the multifarious patterns of life now existing were killed off entirely.

Overpopulated animal colonies cannot be kept alive artificially by providing extra food. Such experiments have been carried out on colonies of rats and mice. After a short while, aggression developed and began to show itself in brutalizing patterns of behaviour and it was posited that fertility declined as a result of unprecedented stress. It is not unreasonable to suppose that our own overcrowded conditions have something to do with our high crime rates.

The problem of overpopulation can be understood relatively easily but it is much more difficult and uncomfortable to have to come to terms with necessary restrictions on material growth. However, it must be made quite clear to every thinking person that limitless expansion of the number of people on this earth and their material needs is impossible. Our present concept of economics is still based on the theory of the ancients that the earth was flat and infinite. However, the rather more limited, globe shaped earth will soon demand an end to all that—if it is not already too late!

There is no doubt whatsoever that demand cannot be allowed to carry on growing, and those who wish to conserve life on earth will have to opt for more modest requirements and for material restrictions. Professor Dr Tschumi, a biologist at the University of Berne, has this to say (119): "As long as demographic and economic expansion continue to be regarded as inevitable or even necessary, without any thought being given to their effects on man and nature, then a catastrophe will be quite unavoidable and any measures that are taken will only serve to deal with certain symptoms in the short term."

On the other hand, those who have learnt a little about futurology (a young science) will find no difficulty in understanding the necessity of putting a rapid end to growth. Futurological theories are often immediately associated in people's minds with concepts like prophecy, astrology and fortune telling, all of which are at once dismissed as nonsense. Those who dismiss such things tend to forget that the most dangerous of all present speculations is perpetrated by those who do nothing, believe that everything will carry on just as before and hope that somehow it will all work out in the end. It would be better, in fact today it is absolutely essential, to calculate in advance just what could be statistically possible if the present rate of growth continues. The earlier such problems are tackled, the more alternatives there will be on which to base decisions. Futurology actually concerns itself, not only with short-term surveys and predictions, but also with the effects of further uncontrolled growth on the next generation and their descendants.

Dr E. Basler, sometime visiting professor at Massachussets Institute of Technology (MIT) in Boston has collected telling evidence in this field (10). He explains that with the maintenance of a mean economic growth rate of only 3·3 per cent a year for the next sixteen years, the environmental changes brought about would equal all those which have occurred in the last 40 years, which would be quite atrocious! In 1985, the numbers of scientists and engineers would have doubled and in a hundred years time, the whole of the natural land surface would be completely built over! The basic elements of our existence would naturally have been destroyed very much earlier. An end to sources of raw materials can also be foreseen.

According to Basler (10), it can also be calculated that mankind today is only 50–100 years, that is two to four generations, away from an absolute turning point, after which the absolute growth rate will have to fall, in order to gradually achieve a state of equilibrium again. This turning point can only be reached by reducing the yearly percentage growth rate until that time. According to Professor Tschumi, whether or not man finds a way back will be decided within a generation (25 years) (119).

By carrying on the present despoilation of the capital of our natural resources, we will only have very few buffer zones and reserves left by the time we achieve the above mentioned state of equilibrium in the biosphere. Mineral resources are almost exhausted now and all we can do is claim interest from our environment. The earth must therefore come to be regarded as quickly as possible as a closed cycle, in which economic limits are imposed by ecological factors. Our consumption must be brought into line with the most economical biological and industrial cycle possible, in which both the flow of artificial energy and the size of the population will be limited.

It is encouraging to note that there are some large-scale industries which have the necessary foresight. Dr Kurt Hess, Chairman of the Rieter machine factory in Winterthur (Switzerland), stated at a press conference (91) that it was industry's task to re-channel expansion to optimum rather than maximum goals—which was taken to mean a conscious renunciation of a purely materialistic concept of economic gain. A few years ago such comments from industrial circles would have been quite unthinkable. Man must therefore learn to renounce materialism. Will he be able to do it?

Basler thinks (10): "that a future without material growth does not necessarily imply stagnation, decadence and tedium, for growth does not correlate with intellectual sciences, artistic or sporting activities but only with material goods. In a cyclic system which was no longer growing, our earth would have practically unlimited potential for all these activities, for which we can spare precious little time in the present scramble for so-called progress."

A fundamental re-orientation of human thinking and education should not be shirked. The present day, purely materialistic ego-mentality will have to be replaced with the notion that, if man wishes to survive, he will have to see himself as a biological being and fit into a natural system which has been wonderfully organized since the creation, and into all its ecological laws.

Professor Dr Jaag, former Director of the Federal Bureau for Water Provision at the ETH in Zurich and an internationally recognized authority in the field of water protection, characterized the present situation most courageously in his farewell lecture, entitled "Must mankind really be destroyed?", in which he said (73): "In order to save mankind from ruin, we have such enormous

164

problems to solve that we shall have to pool all our intellectual and moral resources and try to head off the catastrophe. Up to the present time, men have obviously been basing their undertakings on considerations of war technology, socio-economics and politics, without ever stopping to ask whether their projects were just, responsible or reasonable."

IV · Aspects of the energy crisis

Let us return to our worthy official in the Ministry of Power, who only sees his immediate and restricted task of creating energy at all costs, without considering whether it all makes sense. He simply calculates that the need for energy doubles in so and so many years, as if the world were infinite, and almost certainly believes he is fulfilling a socially useful task.

However, there can be little doubt that a world without nuclear energy would be unlikely to suffer economic disaster. With the technological potential at our disposal today different ways of providing energy could be more intensively researched and ultimately adopted. For example, water power, in an international perspective, is by no means exhausted; and energy problems of the future will increasingly have to be solved at an international level. In pursuing this end, economic problems (like the question of transport) should not be thought of as obstacles, as long as the conservation of life and the environment is being promoted. In this particular area of human relationships, large-scale, international thinking and planning will obviously become increasingly inevitable.

The great danger of nuclear energy, apart from its deadly contamination of the environment, is that it promises man unlimited growth at the very time that he

166

urgently needs to put exactly opposite measures into operation, in order to be able to survive. It is treacherous to paint a picture of nuclear energy as the saviour of mankind arriving just in the nick of time. It merely leads to further lavish wastes of energy and to measureless expansion. Superfluous neon advertisements designed to create unreal needs are well enough known to us all. They are allowed to burn on, year in and year out, prodding consumption to stimulate economic growth, against all the ecological laws and biological requirements, and thus creating an even greater demand for power.

Even at the present time man is using ten to twenty times more energy for combustion than he needs, and in industrial plants, as much as fifty to one hundred times more (110). Moreover, every additional amount of energy, whether it is in the form of heat or nutrients, further weakens the natural ecosystems and ultimately causes environmental pollution. Professor Dr Werner Stumm, Director of the Federal Bureau for Water Provision at the ETH in Zurich, posed the following question in his lecture at the Symposium on Environmental Pollution (110): "From the potential nuclear energy present in rocks, in granite for example, we could release without further ado many times more energy than is released through photosynthesis.[1] To what extent can man, as a geological-biological manipulator, increase the demand for energy and accelerate hydro-geochemical cycles without lowering the stability of the biosphere to such a degree that climatic fluctuations and other evolutionary disasters

[1] In photosynthesis the green plant absorbs carbon dioxide (carbonic acid) and water and builds up from them its own energizing substances with the help of solar energy. Oxygen is released at the same time. Only the cells of green plants can collect light energy from photosynthesis. All other organisms absorb energy in the form of nutrients (carbohydrates, fats proteins).

167

would occur and threaten both the security and further growth of human life?"

An unreasonable demand for energy imperils our health and the basic essentials of life. Indeed material progress at the expense of our health and environment has already reached such a pitch that life could soon no longer be worth living. In the foreseeable future we may possibly have all the material comforts we could desire, but will probably then be forced to admit that those things we were unaccustomed to taking for granted will finally have become unattainable. It is conceivable that, in the near future, the effective standard of living or the *effective state of well-being* will actually fall, even if the *so-called real national income* rises absolutely per capita of the population. Professor Dr H. C. Binswanger, St Gallen College of Economics and Sociology, presented the case for this in 1968. He said (14): "After a slow rise of the so-called real national income and an equally slow retrenchment of the resources at hand (natural raw materials), we have set in motion since the eighteenth and nineteenth centuries, and

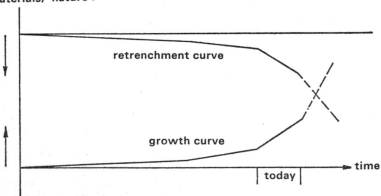

Retrenchment of raw materials, 'nature'.

retrenchment curve

growth curve

today

time

Increase in the so-called real national income

Fig. 7.

with a vengeance from the middle of the twentieth, a process of acceleration which could soon lead to the crossing of the growth and economy curves. From that time onwards the increase in the standard of living caused by growth will be smaller than the drop in the standard of living caused by the despoilation of nature. *Economic growth does not therefore represent an exercise in maximization—which is still accepted as sacrosanct the world over—but rather a problem of optimization!"*

We have already disqualified the underdeveloped

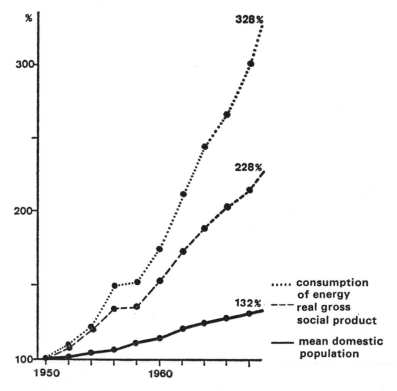

Fig. 8. Percentage increase in Swiss population in real gross social product (in 1958 prices) and in total consumption of energy from 1950 to 1969. Appreciation from 1950=100 per cent. (According to the statistics of the Federal Bureau for Economic Energy).

nations from ever reaching our standard of living, because nature would no longer be able to stand the strain, and the sources of raw materials would be exhausted (10). We are going to have to go without a good deal if this problem is really to be solved. Instead of which, we carry on blithely indulging ourselves in our material culture. The West's consumption of energy, rising senselessly in response to population growth, provides ample evidence of this. As an example, Switzerland can be represented as fig 8, according to Professor Tschumi (118).

Werner Stumm, Professor of the Conservation of Waterways at the ETH in Zurich, states (111): "that the energy used by civilization and the production of entropy are already (1971) too great for our ecosphere to be able to remain in an unimpaired, stable state." Professor Tschumi Professor of Environmental Biology at the University of Berne, thinks (118): "The production of energy is also achieved largely at the expense of our environment, be it as a result of the building of reservoirs for water power stations (loss of recreation space), through the burning of fuels in heating installations, through petrol and diesel motors of the kind used in thermal power stations (air pollution through sulphur dioxide etc, concentrations of carbon dioxide, lack of oxygen), or through atomic power stations (thermal water pollution, danger of radioactive contamination)."

There is certainly an adequate amount of energy for mankind today and the possibilities also exist for *moderated*, reasonable productions of energy for the future. The far-sighted person, who takes note of the ecological laws, will therefore never ask how ever-increasing amounts of energy can be achieved. *The primary question is rather*: *how can we survive?* The demand for energy is only a secondary and relatively minor consideration.

Given our present, environmentally hostile market economies, the supply is tailored to fit the demand. Along

170

with population expansion and a whole series of other factors, these will lead to more and more economic growth —which has so far held an undisputed position in our economic dogma—and to increased energy consumption. It will also lead inevitably to the despoilation of nature.

However, even a planned economy will not be able to deal with a shortage of natural resources. With this method, instead of having the pressure of competition, we have the obligation to fulfil the goals of ever expanding planning, which also implies a necessary and systematic despoilation of natural treasures.

National economies therefore face new tasks. The ultimate aim of a new economic structure would be to take responsible account of the finite nature of our environment. According to Professor Dr H. C Binswanger from St Gallen (15), a new, environmentally-orientated economic system should be sought, in which the demand is dependent on the ecological supply. This would mean that environmental goods would have to be exactly accounted for and the supply limited by ecological possibilities, which would simultaneously establish the economic limits. With this system the restricting factor would be the permissible consumption of energy, which would be reduced to correspond to the ecological equilibrium.

Even today the amount of artificially produced energy will soon reach the level of that produced by photosynthesis. In other words, we are reaching the order of magnitude of natural cycles (120), which is threatening the stability of our biosphere (110). Any further increases in energy consumption should therefore be checked and research should be devoted entirely and exclusively to the development of environmentally favourable forms of energy production.

Under certain circumstances nuclear fusion might present such a possibility. Renowned physicists have calculated a development period of only 20–30 years before this could become commercially operational. Every process possible in physics must also be realized technologically. However, fusion energy does not appear to be entirely free from problems, although, in comparison with energy produced through nuclear fission, it only produces a tiny fraction of the radioactivity. The potential dangers are also very small. Hannes Alfven, Professor of Nuclear Physics and Plasma Research at the University of Stockholm, and Nobel Prize Winner for Physics in 1970, writes of the Swedish atomic energy policy in a letter of 23 August 1970 to the Swedish state council (4): "The process of producing energy from fusion only gives rise to helium as a waste product. One non-radioactive and poison-free noble gas, one radioactive bi-product (tritium) in small quantities and neutron radiation . . . all cause certain problems, which are probably not very serious but which must nonetheless be researched."

Optimism was also in evidence at the fourth International Conference on the peaceful application of nuclear energy, held in Geneva in September 1971. The Max Planck Institute in Garching, near Munich, calculates that the first 1,000 megawatt fusion reactor will be operational by the year 2000 (31). Fusion would provide mankind with an almost inexhaustible supply of energy. It is based on heavy hydrogen and tritium, which is produced from lithium in the reactor.

This implies a re-orientation of the present commercial application of nuclear fission, which above all means experimentation with the second generation of nuclear power stations, the *breeder reactors*. In a purely technological sense, the idea is both ingenious and fascinating, although it does show something of the hideousness of ushering in, by industrial means, a blinded society to whom biological thinking is quite foreign.

In the breeder reactors, the energy-yielding fission material is produced from non-fissile uranium 238 (which

172

appears in nature side by side with fissile uranium 235), in the form of plutonium 239. This type of reactor produces more fission material (plutonium) than it uses! The world supplies of uranium would therefore last for at least 1,000 years, whereas the fuel uranium 235 of the present-day light water reactors is only exploited up to 2–3 per cent and would be exhausted in 30–40 years.

Plutonium is incomparably dangerous; it ignites in the air, which gives rise to the formation of many tiny particles of plutonium oxide. These *"hot particles"* remain in the lungs for a long time if inhaled and can cause lung cancer through the extraordinary concentrations of local alpha-radiation which they produce. Just 1 millionth of a gramme of plutonium injected under the skin of a mouse can give rise to malignancy (35). Plutonium 239 has a half life of 24,000 years and therefore, measured in human terms, can never be destroyed. It is estimated that 30 tons of plutonium a year will be produced by 1980 and by the year 2000, as much as 1,000 tons (35).

The present radiological protection laws offer inadequate safeguards against the dangers of plutonium. The threshold doses recommended by the ICRP are related to an equal dissemination of plutonium in the whole of the lung and not to hot particles. Maximum permissible doses were nevertheless established for the concentration of plutonium oxide in the air by the American Atomic Energy Commission (USAEC) (39). However, there is evidence to show that unequal irradiation of the lungs with radioactive particles can give rise to malignancy before equal irradiation does (35). Research workers have even gone so far as to mention an effect which is a hundred times more powerful (39). The ICRP also stated back in 1965 that this problem could not be satisfactorily assessed (47).

The disastrous prospect of this whole mis-development can also be measured against the fact that the fuel slug in breeder reactors could be increased in size to as much as

1,000 megawatts on economic grounds. Moreover the cooling material used is the highly dangerous natrium, which reacts extremely violently with humidity in the air and with any kind of water. The danger of accidents and harmful side-effects, particularly as a result of catastrophes, would be eliminated many times over in comparison with the already dangerous, present day reactors.

Furthermore plutonium would become available the world over, so that "atom bombs for everybody" would become a reality. It is high time that nations began demanding a total ban on the production of plutonium of any kind, and particularly for the "peaceful application of nuclear energy". The development of breeder reactors must be halted. Even the use of small amounts of plutonium 238 for the production of energy or to provide power in satellites or for space travel is extremely dubious. For example, in 1964 SNAP-9A (Space Nuclear Auxiliary Power = SNAP) burnt up in the atmosphere together with its satellite, with the result that hot particles of plutonium oxide were disseminated over the earth's surface and inhaled by human beings (39). A later security analysis of space travel showed that an accident with SNAP-19, which was to be launched with a weather satellite, would be estimated to cause some 40,000 deadly cases of cancer if burning took place in the atmosphere (39). Apollo 13 (the unsuccessful moon landing) contained SNAP-27 in the LEM. As is well-known the LEM then burnt out on the return journey, together with the SNAP-27 of course. How many people will have to pay for this mishap with a deadly case of lung cancer?

The whole large-scale, technological application of nuclear energy is a fatal blind alley for our civilization. The danger of contamination by the radioactive substances it produces is too great for it to be mutely accepted. New technology must not be allowed to harm the basic elements of human life. Dr Manstein writes with

174

full justification (82): "What use is it holding great congresses about natural, healthy nutrition if, for example, radioactivity is going to become most highly concentrated on the valuable buds of cereals? What value are the long-overdue laws requiring producers to indicate artificial additives in foodstuffs, if, in the meantime, 'anonymous' additives are multiplying and becoming so dangerous as a result of untested methods of fertilization, particularly through the inclusion of radioactive substances? All these important contributory problems are overshadowed by radioactivity, which is rapidly gaining ground in our environment. This has brought about the final and crucial phase in a general, concerted onslaught on biological and genetic substances—although a lot of people have yet to realize it."

Many well meaning people have not noticed it simply because the individual deaths of the banner headlines claim more attention and can be easily understood by everybody. But no attention is paid to man, steering himself to his collective end with official permission and under state protection. It seems that such an atrocity is beyond the conception of most people, for the individual always feels well protected by the mass. Enlightenment through education and the dissemination of frank information is therefore essential.

G•

V · Who has the responsibility for nuclear power stations?

Unfortunately government authorities, businessmen in the atomic industry, and scientists involved, cannot generally be absolved from the accusation that they do not inform the public adequately of the dangers of radioactivity from nuclear power stations. An illustration of this is the following, typical, example.

In Switzerland, more nuclear power stations are at the planning stage, and the communities involved can make their position clear on the projects. To help them do this, the nuclear power industry has up till now carried out broad-based "explanation campaigns" for the public and the local authorities, and have operated these campaigns with up-to-date advertising techniques. These have included newspaper campaigns, exhibitions, and public meetings with lectures and expert advisors, who were labelled "neutral" or "objective", but who actually comprised supporters of atomic industry. Lectures never mentioned the opposing viewpoint. Public debates were as far as possible eliminated or directly avoided, although the customary political practice is to propose and listen to both sides of the argument. An unbiased formation of opinion can only be achieved in this way. In the field of radioactive dangers, quite different practices are the order of the day!

The Gösgen nuclear power station, near Olten in

Switzerland, is one that has received thorough "discussion". It will be situated in a densely populated area of some 100,000 inhabitants who live within a radius of six miles. The regional planning authority of Olten-Gösgen-Gau, whose position in the matter was critical, therefore called upon the experts for detailed information. What happened next was most disturbing.

The authorities' attitude to the dangers of radioactivity is contained in the following quotation (93): "The explanations of objective experts have completely dispelled any concern there may have been at the beginning .... Because the plant can produce no external radiation, a fact supported by the safety regulations, the population density in the vicinity of an atomic power station can be of no crucial significance."

This primitive logic and the erroneous statement about external radiation form a stark contrast to a passage from Professor Tsivoglou's verdict on the projected atomic power station in Kaiseraugst, near Basel (121): "The multiplicity of radioactive substances released is very great and, according to the composition of the waste and the method of processing it, radiation effects in human beings can occur not only directly, but also through a complicated, but actually equally serious environmental cycle. It must therefore be considered impossible to predict with any degree of certainty all the possible inferences that could be made—particularly before a nuclear power station has even become operational." Further comment on the erroneous statements of the regional planning commission mentioned above should be superfluous.

In no way superfluous, however, is the question of who should be held responsible for errors of judgement that could later have decisive consequences. Politicians do not generally have detailed knowledge of the problems of technology and the radioactive dangers of nuclear power stations, so interested parties do not have too much difficulty in getting their own way. Moreover, opponents are usually told that they are isolated, or health maniacs, incompetent, or backward-looking when compared with the "conscientious experts". Besides which, there is rarely a comprehensive explanation of the opposing point of view presented in the mass media.

Flight in the face of actual responsibility is therefore the easiest position for the politician, particularly as he must appear fully justified because all, or almost all, the state experts are supporters of nuclear energy. Characteristic of this position was a leader article in the *Solothurner Az* of 13 February 1971—the area where the 660 megawatt Gösgen atomic power station will be situated—under the title "Democracy looks to Further Education". It reads: "Explanatory articles appear in the newspapers, orientation sessions are held and literature distributed. It is impressive to see a large region taking the trouble to form something like its own opinion in an extremely technical, specialist field. In reality we do not want to be duped, but ultimately we have no choice but to believe the scientists and the specialists. We must trust what they say. Neither the minister, nor the party politician nor the local councillor carries responsibility for such decisions in reality. This lies with the experts and in future it is the experts who will have to be held responsible for errors of judgement. If the local councillors and higher authorities were made responsible, this would be placing exorbitant demands on democracy."

But the matter is not as simple as the writer of this article thinks. People who do not take the trouble to listen to the well-founded arguments of those who warn against atomic power stations, or people who consciously conceal this viewpoint from the public, must take full responsibility along with the experts who advocate the decisions. Every politician, every local community leader who does

178

not seek discussion with the opposing point of view should have an uneasy conscience. It seems equally grotesque that the above quoted newspaper should wish to absolve the government from all responsibility.

However, even the scientists and experts who advocate atomic power stations do not seem to be getting along too well with their bold policies, for growing mutterings in the ranks are seeking expression. Professor Dr W. Winkler, President of the Swiss Atomic Energy Commission, writes in the *Bund* of 25 January 1971, under the title "Nuclear Power Our Cleanest Source of Energy", amongst other things: "The technologist living in an industrial area also recognizes from another viewpoint that, implicit in the course of technological development, there are signs which fill him with great concern. Alvin Toffler writes of this in his most readable book on future shock: 'We are experiencing here the first manifestations of an international revolt, which will shake parliaments in the coming years. This protest against the devastation caused by irresponsibly applied technology could take pathological forms—something like a fascist enemy of the future, which sends scientists to concentration camps instead of Jews. Sick societies need scapegoats....'" (143)

Such observations indicate a possible eventual twist of justice which would call irresponsible scientists to account and which could be attributed *a priori* to purely pathological motives. Even that would be quite an easy way out. Ultimately nobody wants to be held responsible, neither politicians nor scientists, and the public at large, which asks no questions and is never fully orientated, will have to bear the brunt.

Even the most recent and the most powerful committees in the field of radiology, the UNO commission and particularly the ICRP, carry no responsibility. These two scientific bodies, however, are honest in providing the evidence which is lacking.

Alvin Toffler's pertinent formula also applies to the choice of sources of energy (116): "Increasingly, newer and newer and ever more varied innovations will be presented to society and the problem of selection will become more and more complex. The naive old policy of allowing short-term, economic goals to predetermine choice is becoming dangerous and could have catastrophic effects on society." People today must (according to Toffler) make super-decisions, which is why technological questions must not just be answered technologically. "They are political questions and they concern us far more than the other superficial political problems with which we occupy ourselves today."

Frank, open and comprehensive information for the public has therefore become even more necessary than ever, before new technologies are unleashed on society. Before projects get under way—be they new sources of energy, new chemicals, new raw materials—the consequences for our health, for the ecological equilibrium and for the whole of society in the long term must be explained precisely and in detail. A new technology, like the production of energy from nuclear fission, which not only encourages related exponential economic growth but also produces indestructible, cumulative, genetically damaging and deadly poisonous radionuclides in vast quantities and allows them to be released into the environment, really is crazy.

The real orientation of the public must be neglected no longer, not even under the pretext that the public might be shocked and might not be mature enough for the news!

In 1971, Switzerland was the first nation to include environmental protection in the federal constitution, through a series of supplementary legal clause (24 septies).

180

Earlier penal law had proved inadequate to deal with the pollution of air and water and to carry through the battle against noise.

Earlier protection measures against radioactive discharge were based on civil, or (German/Austrian) neighbourhood law, so that an injured person had to file his own suit. In civil law, grievances against radioactive discharges assume either evidence of direct damage or of imminent damage. Such evidence could only be brought by a private citizen at great cost and might not even be able to be proved at all, because damage to people, animals and plants (for example in the normal operation of a nuclear power station) through radioactive discharges has a long-term and not an immediate effect. The latency period for the development of malignant growths for humans, as a result of the deadly effects of very small doses of carcinogenous hydrocarbons (even in the air), can be established as a period of at least 25–30 years (77). Moreover discharges from aeroplanes and cars, etc., cannot be assessed.

The established norms of every country's legal systems are quite ineffectual in terms of environmental protection. Even the classic rights of property in criminal law begin with the premise of a direct connection between the culprit and the victim, and disregard the case of the wrongdoer being anonymous and the victim only statistically identifiable, as would be the case with increased numbers of diseases as a result of environmental pollution (138) or long latency periods.

The protection of human beings and their environment from harmful emissions of radioactivity should be constitutionally guaranteed everywhere in the future. Dr H. E. Vogel, from Zurich, writes on this topic (138), that the new clause in the Swiss constitution "seeks to express that man is not only the master of the natural environment which he has subjected to his will, but that he also has functions to fulfil in the capacity of a trustee. He is therefore responsible and limited in his rights of decree, not only in relation to living fellow human beings, but also to future generations and the other forms of life, animals and plants which belong to his natural environment. . . . The criterion for protective precautions must under no circumstances be defined primarily by economic considerations or ventures. Industrial concerns whose profitability depends on the pollution of air or water or on the creation of excessive noise, must be seen as intolerable parasites even from the point of view of economics." This applies to the whole process, beginning at the uranium mine and ending with atomic waste.

In relation to the dangers of radiation inherent in the peaceful application of nuclear energy, the ICRP states quite clearly, *that it is not yet possible to balance the risks and benefits, since it requires a more accurate quantitative appraisal of both the possible biological damage and the probable benefits than is now possible* (48).

Nevertheless we blithely carry on contaminating our environment by allowing the release of demonstrably mutagenous substances from nuclear power stations, which can cause deadly diseases. Man, as a trustee for future generations, is handling his precious genotype quite irresponsibly, and it is hardly surprising that the top international scientific bodies in the field of radiobiolgy should disclaim practically all responsibility. UNSCEAR writes: (137): "An additional difficulty with genetic risks is encountered in expressing them in meaningful terms . . . . The majority of genetic injuries cannot be assessed and indeed only in the minority of cases can we make assumptions as to the manner in which the damage will be displayed in future generations, for the individual and for society."

The ICRP thinks (50): "Because the total genetic damage will become manifest only over very many

182

generations, it is appropriate that long-term effects should be a major preoccupation of the collective conscience; from this viewpoint the 'total eternity damage' can be considered." This shows quite clearly, that science today is in no position to calculate in advance the total risks of the application of nuclear energy for peaceful purposes.

The hackneyed argument that the nuclear industry is the safest industry of all because it causes so few accidental deaths and injuries is misleading and designed to deceive a gullible public. The gain from present nuclear power stations in terms of the production of electricity is immediately apparent, but the pay-off in terms of harm will last far into the future, for years and decades. Once caused, genetic damage will be irreversible—even if it is visually or statistically recognizable—for millions of people will, by then, have received high radiation doses which can never be reversed.

We are already having to pay for many of the erroneous assessments of the unchecked developments of civilization, which have continued unheeded under the eyes of the scientists and the authorities. General environmental pollution has become evident to everybody now. One independent science recognized the danger a long time ago but technology and industry chose to disregard the warning voices and to concentrate solely on material profit. Even dependent science ignored all warnings and chose instead to laugh them off. However it would be unfair to make industry the only scapegoat for the present situation, for it is not only the nuclear industry which must be re-orientated now: *the fundamental habits of all human beings must change. This implies far-reaching changes for every single household.*

Moreover, economic policy must be expanded to embrace new aims. Previously people strived exclusively for ever-increasing rises in the so-called national income and in the social product, without paying the slightest

183

attention to the despoilation of our natural resources and the concomitant threat to our health and existence. In future technology must not be allowed to impair the quality of life and the environment.

In the summer of 1971, 2,100 biologists and environmental research workers from twenty-three countries signed a petition prepared by the UN General Secretary, U Thant (8), which warns not only against environmental damage and other threats to human beings, like air and water pollution, over-population and world-wide aggression, but also demands a deferment of plans for the creation of more nuclear power stations. The appeal to U Thant for this memorandum was made by four Nobel Prize winners and other prominent scientists from a number of different fields. Apart from this, over 100 scientists from twelve nations met in Trondheim (Norway) on 14 October 1971, in the context of an international conference of The Society for Social Responsibility in Science. They directed an appeal to all scientists to cease taking part in the construction of nuclear power stations and demanded an end to the building of such reactors (22a).

The problem of nuclear power stations should not be judged by experts and biased scientists alone. The ICRP draws attention to this in the last remark quoted. It is therefore high time that the public was enlightened about the dangers threatening mankind through nuclear fission. Each and every one of us has an essential right to participate in the decision.

References

Abbreviations:
EAWAG   Eidgenössische Anstalt für Wasserversorgung
ETH   Eidgenössische Technische Hochschule (Zürich)
ICRP   (Publications of) International Commission on
             Radiological Protection. Pergamon Press Ltd.,
             Oxford
SVA   Schweizerische Vereinigung für Atomenergie
UNSCEAR (Reports of the) United Nations Scientific Com-
             mittee on the Effects of Atomic Radiation.
             United Nations, New York
WSL   Weltbund zum Schutze des Lebens

1. Prof. Dr. H. Aebi, Rektor der Universität Bern, Symposium über den "Schutz unseres Lebensraumes", 10/12 Nov. 1970, ETH, Zürich.
2. Dr. F. Alder, "Sicherheit und Risiko von Kernkraftwerken". Informationstagung der SVA, 4/6 Nov. 1970 in Bern. Separatdruck, S.48.
3. Dr. F. Alder, "Sicherheit und Risiko von Kernkraftwerken". Informationstagung der SVA, 4/6 Nov. 1970 in Bern. Separatdruck, S.49.
4. Prof. Dr. H. Alfven, TH Stockholm, Brief vom 27.8.70. Übersetzung veröffentlicht durch Walter Soyka, Gesellschaft für biologische Sicherheit, Wien, 1971.
5. Dr. H. Ambühl, ETH, EAWAG, Zürich, Informationsblatt Nr. 17, 1970 der Föderation Europäischer Gewässerschutz, S. 15.

6. Ärztememorandum gegen Kernspaltungs-Kraftwerke. Österreichische Ärztezeitung, Nr. 20, 25. Okt. 1970.
7. "Atommüll in den Atlantik". Hamburger Abendblatt, 31 July 1971.
8. Atomwirtschaft Atomtechnik, Heft 6, 1971, S. 270.
9. Ing. J. Bächtold, "Umweltgestaltung". Symposium über den "Schutz unseres Lebensraumes", 10/12 Nov. 1970, ETH, Zürich.
10. Dr. Ing. E. Basler, Zürich, Guest Professor at Massachussets Institute of Technology (MIT), Boston, Abschiedsvorlesung 1970. Tages-Anzeiger (Zürich), 5 Dec. 1970 (Magazin-Beilage).
11. Prof. Dr. K. Bechert, Die Bundesregierung informiert. Das Gewissen. Zeitschrift für Lebensschutz, Nr. 2, Hilchenbach, 1969.
12. Dr. med. F. Becker, Moderne Krankheitsbehandlung oder gesunde Lebensführung. Sonderdruck. Verlagsgenossenschaft der Waerland-Bewegung, Mannheim, 1964.
13. J. Binder, Nationalrat, Baden, Es geht um den Menschen. Wasser-Boden-Luft (Verlag A. Grob, St. Gallen), Herbstausgabe 1970, S. 24–26.
14. Prof. Dr. H. C. Binswanger, Hochschule St. Gallen für Wirtschaftsund Sozialwissenschaften, Antrittsvorlesung. St. Gallen, 1968.
15. Prof. Dr. H. C. Binswanger, "Eine umweltskonforme Wirtschaftsordnung?" Symposium für wirtschaftliche und rechtliche Fragen des Umweltschutzes, Hochschule St. Gallen, Okt. 1971.
16. Dr. M. Bircher-Benner, Diätbücher. Bircher-Benner Verlag, Erlenbach.
17. Dr. R. Bircher, Umsturz in der Ernährungslehre. Der Wendepunkt, Nr. 4, 1970, S. 145–149.
18. Prof. Dr. R. Braun, ETH, EAWAG, Zürich, Vortrag im Rotary Club Lenzburg, 9 Nov. 1970.
19. Dr. med. M. O. Bruker, Aus der Sprechstunde. Buchreihe "Schlank ohne zu hungern durch vitalstoffreiche Ernährung", Band 3. W. Schnitzer Verlag, St. Georgen/ Schwarzwald, 1969.
20. Dr. med. M. O. Bruker, Was von den Befürwortern der

Atomkraftwerke verschwiegen wird. WSL, Arbeitskreis Atom, Bad Godesberg, 1970.

21. Dr. H. Brunner, Die ICRP steht nicht still. SVA-Bulletin Nr. 3, 1971.

22. R. Curtis and E. Hogan, Das Märchen vom friedlichen Atom. Schriftenreihe "glücklicher leben", Nr. 11, Salzburg, 1970.

22a. Das Gewissen. Zeitschrift für Lebensschutz, Nr. 11, Hilchenbach, 1971.

23. Das sagen die Atomexperten. Fortschritt für alle, Sonderheft (Verlag Gemeinnützige Aktion Fortschritt für alle, Feucht), 1971, S. 38/39.

24. Prof. Dr. M. H. DeGroot, Prof. of Mathematical Statistics and Head of the Department of Statistics at Carnegie-Mellon University, Pittsburg. Brief on Senator Edwin G. Hall, State Capitol Building, Harrisburg, Pennsylvania, 20 Okt. 1970.

25. "Demokratie wird zur Volkshochschule". Solothurner AZ, 13 Febr. 1971.

26. Dr. med. R. Drobil, Die Menschenbombe. Schriftenreihe "glücklicher leben", Salzburg, 1968.

27. Doppelstab (Basel), 29 Juni 1970.

28. Eidgenössisches Departement des Innorn, Richtlimen über die Beschaffenhen abzuleitender Abwässer, Vom 1 Sept. 1966. Drucksachenkanzlei des Bundeshauses, Bern, 1966.

29. "Erwiesener Kindermord". Gesundes Leben. Nr. 22, Linz, 1970.

30. "Europas größtes Atomkraftwerk Biblis". Neue Zürcher Zeitung, Nr. 133, 1971.

31. Festschrift 1970. Max Planck-Institut für Plasmaforschung, Garching.

32. Dr. P. Feuz, SVA, "Keine Gefahr durch Atomkraftwerke". Neue Zürcher Zeitung, Nr. 521, 1969.

33. Prof. Dr. H. Fritz-Niggli, Universität Zürich, "Strahlenwirkungen auf biologische Objekte, insbesondere den Menschen". Informationstagung der SVA, 4/6 Nov. 1970 in Bern. Separatdruck, S. 31 und 32.

34. Prof. Dr. H. Fritz-Niggli, "Strahlenbiologie eine Wissenschaft". Basler Nachrichten, 1 Juli 1970.
35. D. P. Geesaman, Lawrence Strahlenlaboratorium, Kalifornien. "Plutonium and Public Health". Symposium at the University of Colorado, 19 April 1970.
36. Dr. J. W. Gofman and Dr. A. R. Tamplin, Bericht für den Unterausschux über Luft- und Wasserverschmutzung für öffentliche Arbeiten. Senat der USA, 91 Kongreß, 18 Nov. 1969.
37. Dr. J. W. Gofman and Dr. A. R. Tamplin, "Ein Kongreß-Seminar", 7/8 April 1970, Strahlenlaboratorium der Universität Berkeley.
38. Dr. J. W. Gofman and Dr. A. R. Tamplin, Poisoned Power. The Case against Nuclear Power Plants. Rodale Press, Emmans, Pa., 1971.
39. Dr. J. W. Gofman and Dr. A. R. Tamplin, Population Control through Nuclear Pollution. Nelson-Hall Company, Chicago, 1970.
40. Prof. Dr. H. Grümm, Reaktorzentrum Seibersdorf, "Die Umweltbelastung bei Kernkraftwerken". Informationstagung des österreichischen Atomforums in Wien, 25 März 1971.
41. Prof. Dr. H. Grümm, "Die Sicherheitsbilanz der Reaktoren". Informationstagung der SVA, 4/6 Nov. 1970 in Bern. Separatdruck, S. 63.
42. Dipl. Ing. G. Häringer, Ministerialrat, Düsseldorf, Vom Stand der Rheinwasserverschmutzung. Wasser-Boden-Luft, (Verlag A. Grob, St. Gallen), Frühjahrsausgabe 1970, S. 43.
43. Dr. W. Herbst, Radiologisches Institut der Universität Freiburg i. Br. Österreichische Ärztezeitung, Nr. 20, 25. Okt. 1970, S. 2475–2480.
44. Dr. W. Herbst, "Strahlenschutzpraxis in der Defensive". Vitalstoffkongreß 1970 in Luxemburg, 11. Sitzung, Referat Nr. 72.
45. Dr. W. Herbst, Unübersehbare Gesundheitsrisiken durch Lebensmittelbestrahlung. Verlagsgenossenschaft der Waerland-Bewegung, Mannheim, 1968.
46. Prof. Dr. D. Högger, Präsident der eidgenössischen

Kommission für Lufthygiene, "Wie lange reicht der Sauerstoffvorrat der Atmosphäre?" Neue Zürcher Zeitung, Nr. 115, 1971.
47. ICRP-Publication Nr. 9, 1965, S. 4.
48. ICRP-Publication Nr. 9, 1965, S. 15, Ziff. 81.
49. ICRP-Publication Nr. 9, 1965, S. 15, Ziff. 83.
50. ICRP-Publication Nr. 8, 1966, S. 2.
51. ICRP-Publication Nr. 8, 1966, S. 3.
52. ICRP-Publication Nr. 8, 1966, S. 5.
53. ICRP-Publication Nr. 8, 1966, S. 51.
54. ICRP-Publication Nr. 8, 1966, S. 55/56.
55. ICRP-Publication Nr. 8, 1966, S. 56.
56. ICRP-Publication Nr. 8, 1966, S. 57/58.
57. ICRP-Publication Nr. 8, 1966, S. 59.
58. ICRP-Publication Nr. 8, 1966, S. 60.
59. ICRP-Publication Nr. 14, 1969, S. 5.
60. ICRP-Publication Nr. 14, 1969, S. 10.
61. ICRP-Publication Nr. 14, 1969, S. 11.
62. ICRP-Publication Nr. 14, 1969, S. 13.
63. ICRP-Publication Nr. 14, 1969, S. 14.
64. ICRP-Publication Nr. 14, 1969, S. 22.
65. ICRP-Publication Nr. 14, 1969, S. 23.
66. ICRP-Publication Nr. 14, 1969, S. 28.
67. ICRP-Publication Nr. 14, 1969, S. 31/32.
68. ICRP-Publication Nr. 14, 1969, S. 32/33.
69. ICRP-Publication Nr. 14, 1969, S. 57.
70. ICRP-Publication Nr. 14, 1969, S. 82.
71. ICRP-Publication Nr. 14, 1969, S. 112/113.
72. ICRP-Publication Nr. 15, 1970, S. 3/4.
73. Prof. Dr. O. Jaag, ETH Zürich, Abschiedsvorlesung am ETH-Tag, 14 Dez. 1970, "Muß die Menschheit wirklich zugrunde gehen?"
74. Dr. D. Kistner, Radionuklide und Lebensmittel. Bundesforschungsanstalt für Lebensmittelfrischhaltung, Karlsruhe, 1962.
75. Dr. med. W. Kollath, Die Ordnung unserer Nahrung. Hippokrates-Verlag, Stuttgart, 1960.
76. Dr. K. Kühn, "Die radioaktiven Abfälle". Information-

stagung der SVA, 4/6 Nov. 1970 in Bern. Separatdruck, S. 58.

77. Prof. Dr. W. Kutscher, Über Benzpyren in Ruß und Luft. Gesundheitstechnik (Verlag A. Brunner, Zürich), Nr. 5, 1971, S. 87.

78. W. H. Langham, Radiobiological Factors in Manned Space Flight. National Academy of Sciences, National Research Committee, Washington.

79. W. Leonhard, "Kernenergie am umweltsichersten". Aargauer Tagblatt, Okt. 1970.

80. Dr. K. H. Lindackers u. a., Kernenergie, Nutzen und Risiko. Deutsche Verlags-Anstalt, Stuttgart, 1970.

81. Dr. H. R. Lutz, "Radioaktive Abfälle und ihre Lagerung". Basler Nachrichten, 2 Juli 1970.

82. Dr. B. Manstein, Im Würgegriff des Fortschritts. Europäische Verlagsanstalt, Frankfurt a. Main, 1961.

83. Dr. S. Mauch, Massachusetts Institute of Technology (MIT), Boston, "Gefahren und Grenzen einer stetig wachsenden Weltbevölkerung". Neue Zürcher Zeitung, Nr. 43, 1971.

84. Dr. med. habil. C. E. Mehring, "Immunitätslage der Bevölkerung nach Erhöhung der Umweltradioaktivität". Internationaler Konvent der Gesellschaft zur Erforschung von Zivilisationskrankheiten und Vitalstoffen, Montreux, 12 Sept. 1971.

85. Dr. med. habil. C. E. Mehring Auffälligkeiten im weißen Blutbild der Bevölkerung in neuerer Zeit. Sonderdruck aus Medizinische Welt (Schattauer Verlag Stuttgart), 24 Okt. 1970, Nr. 43.

86. Dr. med habil. C. E. Mehring, Statistische Untersuchungen zur Leukozytendepression in den letzten Jahrzehnten. Hippokrates. Jg. 40, Heft 23, 1969, S. 922.

87. Neue Zürcher Zeitung, Nr. 216, 1970.

88. Österreichische Ärztezeitung, Nr. 20, 25 Okt. 1970, S. 2434/35.

89. Prof. Dr. L. Pauling. Genetic and Somatic Effects of High-Energy Radiation. Bulletin of the Atomic Scientists, Sept. 1970, S. 3.

190

90. Pressedienst des Bundesministeriums für Bildung und Wissenschaft, 19 Aug. 1970.

91. "Pressetag der Maschinenindustrie". Aargauer Tagblatt, 12 Dez. 1970.

92. Protokoll der Parlamentarischen Gruppe für Natur- und Heimatschutz bei der Vortrags- und Diskussionstagung über Probleme der Atomkraftwerke, 16 Dez. 1970 im Parlamentsgebäude in Bern.

93. "Regionalplanung opponiert nicht". Aargauer Tagblatt, 9 Febr. 1971.

94. Regionalplanungsgruppe "Olten-Gösgen-Gäu", "Nach bestern Wissen und Gewissen". Oltner Tagblatt, 26 Febr. 1971.

95. Dr. Ing. E. H. Reinhardt, Das Gewissen, Zeitschrift für Lebensschutz, Nr. 10, Hilchenbach, 1970.

96. Dr. M. Ruf, Bayerische biologische Versuchsanstalt, München, Die radioaktive Abfallbeseitigung aus Atomreaktoren in die menschliche Umwelt mit besonderer Berücksichtigung der Gewässer. Zentralblatt für Veterinärmedizin, Beiheft 11, 1970.

97. Dr. M. Ruf, Radiobiologische Untersuchungen über die Konzentration und Verteilung des langlebigen Kernwaffen-Fallout in pflanzlichen und tierischen Organismen sowie in den Grundsedimenten von Oberflächengewässern. Habilitationsschrift, München, 1967.

98. Dr. J. G. Schnitzer, Merkblätter der Aktion Mönchweiler. W. Schnitzer Verlag, St. Georgen/Schwarzwald, 1964, S. 20.

99. Schriftenreihe "glücklicher leben", Salzburg.

100. Dr. Ing. E. Schulz, Vorkommnisse und Strahlenunfälle in kerntechnischen Anlagen. Verlag Karl Thiemig, München, 1965, S. 11.

101. Dr. Ing. E. Schulz, Vorkommnisse und Strahlenunfälle in kerntechnischen Anlagen. Verlag Karl Thiemig, München, 1965, S. 234.

102. G. Schuster, Ministerialdirigent im Bundesministerium für Bildung und Wissenschaft, Bonn. Informationstagung der SVA, 4/6 Nov. 1970 in Bern. Separatdruck, S. 5.

103. Prof. Dr. H. Schweigart, Lebensschutz oder Untergang.

Schriftenreihe "glücklicher leben", Nr. 15, Salzburg, 1970.

104. Dr. D. Spain, Iatrogene Krankheiten. Georg Thieme Verlag, Stuttgart, 1967.

105. Prof. Dr. E. Sternglass, "Infant Mortality and Nuclear Power Generation". Hearing of the Pennsylvania Senate Select Committee on Reactor Siting, Harrisburg, 21 Oct. 1970.

106. Prof. E. J. Sternglass, "Low Level Radiation Effects on Infants and Children in the New York Metropolitan Area", University of Pittsburg, 7 May 1971.

107. Prof. Dr. E. J. Sternglass, "Infant Mortality Changes Near the Big Rock Point Nuclear Power Station, Charlesvoix, Michigan". University of Pittsburg, 6 Jan. 1971.

108. Prof. Dr. E. J. Sternglass, "Infant Mortality Changes Near the Peach Bottom Nuclear Power Station in York County", Pennsylvania. University of Pittsburg, 7 Feb. 1971.

109. Dr. A. Stewart and G. W. Kneale, British Medical Journal, The Lancet, 6 Jun. 1970, S. 1185.

110. Prof. Dr. W. Stumm, ETH Zürich, "Der menschliche Konflikt in Nutzung und Bewahrung der Natur". Symposium über den "Schutz unseres Lebensraumes", 10/12 Nov. 1970, ETH Zürich.

111. Prof. Dr. W. Stumm, "Manipulation der Umwelt durch den Menschen". Neue Zürcher Zeitung, Nr. 441, 1971.

112. SVA, "Kernkraftwerke in entlegenen Gebieten errichten". Aargauer Tagblatt, 24 Mai 1971.

113. Symposium über den "Schutz unseres Lebensraumes", 10/12 Nov. 1970, ETH Zürich, Seminar IV, "Lufthygiene".

114. Prof. Dr. M. Thürkauf, Universität Basel, Atomkraftwerke 50 Jahre zu früh. Doppelstab Verlag, Basel, 1970.

115. Prof. Dr. M. Thürkauf, Wissenschaft schützt vor Torheit nicht. Doppelstab Verlag, Basel, 1970.

116. A. Toffler, Der Zukunftsschock. Scherz Verlag, Bern, 1970.

117. Prof. Dr. P. Tschumi, Universität Bern, Allgemeine Biologie. Verlag Sauerländer, Aarau, 1970.

118. Prof. Dr. P. Tschumi, "Existenzgrundlagen der zivilisierten Menschheit". Neue Zürcher Zeitung, Nr. 400, 1971.
119. Prof. Dr. P. Tschumi, Symposium über den "Schutz unseres Lebensraumes", 10/12 Nov. 1970, ETH, Zürich.
120. Prof. Dr. P. Tschumi, "Umwelt als beschränkender Faktor für Bevölkerung und Wirtschaft". Symposium für rechtliche und wirtschaftliche Fragen des Umweltschutzes, Hochschule St. Gallen, Okt. 1971.
121. Prof. E. Tsivoglou, Kontrolle der radioaktiven Verseuchung in Kaiscraugst. Gutachten ausgearbeitet im Auftrag des Wasserwirtschaftsamtes des Kantons Baselland, 4 Juni. 1970.
122. UNSCEAR, 17. Session, 1962, S. 7, Ziff. 47.
123. UNSCEAR, 17. Session, 1962, S. 7/8, Ziff. 48.
124. UNSCEAR, 17. Session, 1962, S. 10, Ziff. 25.
125. UNSCEAR, 17. Session, 1962, S. 10, Ziff. 30/31.
126. UNSCEAR, 17. Session, 1962, S. 12, Ziff. 52/53.
127. UNSCEAR, 17. Session, 1962, S. 19, Ziff. 58/59.
128 UNSCEAR, 17 Session, 1962, S 20, Ziff 6
129. UNSCEAR, 17. Session, 1962, S. 20, Ziff. 10.
130. UNSCEAR, 17. Session, 1962, S. 21, Tab. I.
131. UNSCEAR, 17. Session, 1962, S. 34, Ziff. 34.
132. UNSCEAR, 17. Session, 1962, S. 35, Ziff. 51/52.
133. UNSCEAR, 17. Session, 1962, S. 145, Ziff. 248.
134. UNSCEAR, 17. Session, 1962, S. 148, Ziff. 274.
135. UNSCEAR, 17. Session, 1962, S. 417, Ziff. 34.
136. UNSCEAR, 19. Session, 1964, S. 7, Ziff. 5.
137. UNSCEAR, 21. Session, 1966, S. 122, Ziff. 247.
138. Dr. H. E. Vogel, Rechtliche Aspekte des Schutzes des Menschen und seiner natürlichen Umwelt in der Schweiz. gwf-wasser-abwasser (Vertag Oldenbourg, München), Nr. 1, 1971, S. 27–39.
139. Are Waerland, Der Weg zu einer neuen Menschheit. Verlagsgenossenschaft des Waerland-Bewegung, Mannheim, o. J.
140. Ebba Waerland, Auszüge aus dem Gutachten von Dr. W. Herbst. Freiburg i. Br., zu den Gesundheitsrisiken atomarer Strahlung unter dem Gesichtspunkt natur-

gemäßer Ernährung. Waerland Monatshefte Nr. 1, Mannheim, 1964.

141. "Wie sicher sind Atomkraftwerke?". Tages-Anzeiger (Zürich), 18 Nov. 1970.

142. Prof. Dr. W. Winkler, "Die Sicherheitsaspekte der Reaktorphysik und die Entstehung der Spaltprodukte". Informationstagung der SVA über die Sicherheit von Kernkraftwerken, 4/6 Nov. 1970 in Bern. Separatdruck, S. 9.

143. Prof. Dr. W. Winkler, "Kernenergie unsere sauberste Energiequelle". Der Bund (Bern), 25 Jan. 1971.

144. Prof. Dr. K. Wuhrmann, ETH Zürich, Vortrag am Symposium über den "Schutz unseres Lebensraumes", 10/12 Nov. 1970, ETH Zürich.

145. E. Zimmerli, Tragt Sorge zur Natur. Verlag Sauerländer, Aarau, 1970.

146. Prof. Dr. h. c. W. Zimmermann, Die Statistiken Sternglass. Offener Brief vom 24 Febr. 1971 an die Abteilung für Wissenschaft und Forschung, Bern.